DAVID BRENNAN

# SUSTAINABLE PROCESS ENGINEERING

## CONCEPTS, STRATEGIES, EVALUATION, AND IMPLEMENTATION

PAN STANFORD PUBLISHING

*Published by*

Pan Stanford Publishing Pte. Ltd.
Penthouse Level, Suntec Tower 3
8 Temasek Boulevard
Singapore 038988

Email: editorial@panstanford.com
Web: www.panstanford.com

**British Library Cataloguing-in-Publication Data**
A catalogue record for this book is available from the British Library.

**Sustainable Process Engineering: Concepts, Strategies, Evaluation, and Implementation**

Copyright © 2013 Pan Stanford Publishing Pte. Ltd.

**Cover photograph**
The front cover shows part of the Origin Energy operated natural gas plant, located in the Otway region of Victoria, Australia. This view of the plant surrounded by pristine countryside is intended to symbolise the potential for process plants and nature to coexist. Permission by Origin Energy to use the plant photograph is gratefully acknowledged.

ISBN 978-981-4316-78-1 (Hardcover)
ISBN 978-981-4364-22-5 (eBook)

Printed in the USA

# Contents

*Acknowledgements*                                                    xvii
*Preface*                                                             xix

PART A: CONCEPTS

**Introduction to Part A**                                              3

**1  Sustainability Concepts**                                          9
  1.1  The Concept of Sustainable Development                  9
  1.2  Sustainability in the Context of the Process Industries 13
  1.3  Some Temporal Characteristics of Sustainability         14
    1.3.1  Time Horizons in Project Evaluation       14
    1.3.2  Time Horizons for Technology Development   14
    1.3.3  Time Dependence of Technology Improvement  14
    1.3.4  Robustness to Technological, Economic, and
        Regulatory Change         15
    1.3.5  Appraisal of Uncertainties (Technical, Business, and
        Environmental)            15
  1.4  The Sustainable Project or Industry                     15
  1.5  Conflicts in Achieving Sustainability Objectives        16

**2  Cleaner Production**                                               19
  2.1  Introduction                                           19
  2.2  The Concept of 'Cleaner Production'                     19
  2.3  The Product Life Cycle                                  20
  2.4  Hierarchy of Waste Management                           23
  2.5  Concepts and Sources of Waste                           24
    2.5.1  Concepts of Waste                         24
    2.5.2  Process and Utility Waste                  24
    2.5.3  Utility Waste and System Boundary Definition 25
    2.5.4  Packaging                                 27

2.6  Impacts of Waste                                           27
2.7  Classification of Waste                                    27
2.8  Driving Forces for Cleaner Production                      28
2.9  Resistances to Introducing Cleaner Production              28
2.10 Concluding Remarks                                         29

**3  Industrial Ecology                                         31**
3.1  The Basic Concept of Industrial Ecology                    31
3.2  Energy and Materials Recovery from Waste Streams           34
3.3  Resource Flow through the Economy                          34
     3.3.1  Sulphur Flow in Australia                           34
3.4  Transport and Storage of Raw Materials and Products        35
     3.4.1  Marine Transport                                    36
     3.4.2  Road and Rail Transport                             36
3.5  Integrated Site Manufacture                                36
3.6  Some Examples of Industrial Ecology Initiatives            38
     3.6.1  Case 1: Hydrogen Utilisation from Refineries        38
     3.6.2  Case 2: Fertiliser Complex, Queensland, Australia   39
     3.6.3  Case 3: Industrial Integration at Kalundborg, Denmark  40
     3.6.4  Case 4: Industrial Symbiosis at Kwinana, Western
            Australia                                           41
3.7  Concluding Remarks                                         42

**Problems:  Part A                                             45**

PART B: STRATEGIES

**Introduction to Part B                                        53**

**4  Waste Minimisation in Reactors                             55**
4.1  Introduction                                               55
4.2  A Checklist for Reaction Systems and Reactors              56
4.3  Chemistry of Process Route                                 57
     4.3.1  Conversion, Selectivity, and Yield                  59
     4.3.2  Co-Product and By-Product Utilisation               60
4.4  Impurities in Reactor Feedstocks                           60
4.5  Mixing of Reactants                                        62
     4.5.1  Mixing of Gaseous Reactants                         62
     4.5.2  Mixing of Liquids                                   62
     4.5.3  Fluid Distribution in Packed Bed Reactors           62

4.6    Minimising Secondary Reactions                              63
4.7    Recycle of Unreacted Feed from Reactor Outlet               64
4.8    Reversible Reactions                                        64
4.9    Catalysis                                                   65
       4.9.1  Example of the Effect of Catalyst Activity on
              Performance                                          66
4.10   Agent Materials                                             67
4.11   Case Examples                                               67
4.12   Chlor-Alkali Production in Mercury Cell                     68
       4.12.1 Transport Paths                                      69
       4.12.2 Other Aspects of the Mercury Cell Chlorine
              Process                                              71
4.13   Ethylene Manufacture from Hydrocarbons                      71
4.14   Hydrogen Cyanide Manufacture from Ammonia, Methane,
       and Air                                                     73
4.15   Sulphuric Acid Manufacture                                  75
4.16   PVC Production by Suspension Polymerisation of Vinyl
       Chloride Monomer                                            78
4.17   Concluding Remarks                                          80

5  **Waste Minimisation in Separation Processes**                  **83**
5.1    Classification of Separation Processes                      83
5.2    Sources of Waste in Separation Processes                    84
5.3    Distillation                                                85
5.4    Gas Absorption                                              87
5.5    Adsorption                                                  90
5.6    Filtration                                                  91
       5.6.1  Centrifugal Separation                               92
       5.6.2  Filtration of Solids from Gas Streams                92
       5.6.3  Separation of Liquid Particulates from Gas
              Streams                                              92
5.7    Drying                                                      93
5.8    Evaporation and Condensation                                93
5.9    Solid–Liquid Extraction                                     94
5.10   Liquid–Liquid Extraction                                    95
5.11   Use of Extraneous Materials                                 95
       5.11.1 Example of Extraneous Material Use — Sulphuric Acid
              in Chlorine Drying                                   96
5.12   Case Examples                                               97
       5.12.1 Case Example — Solid Sodium Cyanide Plant            98

5.12.2 Other Case Examples of Gas Absorption in Chemical
　　　　Processes　　　　　　　　　　　　　　　　　　99
5.12.3 Case Examples in Distillation　　　　　　　　　100
5.13 Concluding Remarks　　　　　　　　　　　　　　101

**6　Identification of Waste in Utility Systems　　　　　103**
6.1　Introduction　　　　　　　　　　　　　　　　　103
6.2　Fuels　　　　　　　　　　　　　　　　　　　　105
6.3　Fuel Combustion　　　　　　　　　　　　　　　105
　　　6.3.1　Heat of Combustion　　　　　　　　　　　106
　　　6.3.2　Excess Air　　　　　　　　　　　　　　　107
6.4　Common Fuels　　　　　　　　　　　　　　　　107
6.5　Environmental Impacts of Flue Gases　　　　　　109
　　　6.5.1　$NO_x$ Formation in Fuel Combustion　　　　110
6.6　Theoretical Flame Temperatures　　　　　　　　111
6.7　Furnaces　　　　　　　　　　　　　　　　　　111
6.8　Flare Stacks　　　　　　　　　　　　　　　　112
6.9　Steam Generation　　　　　　　　　　　　　　113
6.10 Steam Use　　　　　　　　　　　　　　　　　116
6.11 Water Sources and Uses　　　　　　　　　　　119
　　　6.11.1 Water Quality Indicators　　　　　　　　120
6.12 Recirculated Cooling Water from Cooling Towers　121
6.13 Sea Water Cooling　　　　　　　　　　　　　123
6.14 Air Cooling　　　　　　　　　　　　　　　　124
6.15 Refrigeration　　　　　　　　　　　　　　　124
6.16 Electricity Demand and Supply　　　　　　　126
6.17 Distribution and Use of Electricity　　　　　129
6.18 Compressed Air　　　　　　　　　　　　　130
6.19 Inert Gas　　　　　　　　　　　　　　　　130
6.20 Vacuum　　　　　　　　　　　　　　　　131
6.21 Concluding Remarks　　　　　　　　　　　131

**7　Energy Conservation　　　　　　　　　　　　133**
7.1　Introduction　　　　　　　　　　　　　　　133
7.2　Energy Consumption in Compression of Gases　134
　　　7.2.1　Process Specification for Gas Compressors　134
　　　7.2.2　Machine Selection　　　　　　　　　　134
　　　7.2.3　Thermodynamics of Gas Compression　136
　　　7.2.4　Limits to Compression Ratio per Stage of Compression　138
　　　7.2.5　Intercooling of Gas during Compression　138

|        | 7.2.6 | Reliability                                           | 138 |
|--------|-------|------------------------------------------------------|-----|
|        | 7.2.7 | Drives for Compressors                               | 139 |
|        | 7.2.8 | Energy Conservation in Gas Compression               | 139 |
| 7.3    | Energy Consumption in Pumping of Liquids            |      | 139 |
|        | 7.3.1 | Process Specification for Pumps                      | 139 |
|        | 7.3.2 | Power Requirement                                    | 141 |
|        | 7.3.3 | Pump Machine Types                                   | 141 |
|        | 7.3.4 | Centrifugal Pump Selection and Performance           | 142 |
|        | 7.3.5 | Energy Conservation in Pumping of Liquids            | 144 |
| 7.4    | Pressure Losses in Piping                           |      | 144 |
|        | 7.4.1 | Sizing of Pipes                                      | 144 |
| 7.5    | Pressure Loss through Equipment                     |      | 145 |
|        | 7.5.1 | Heat Exchangers                                      | 145 |
|        | 7.5.2 | Vapour–Liquid Contacting Columns                     | 146 |
| 7.6    | Agitation and Mixing                                |      | 147 |
| 7.7    | Heat Recovery                                       |      | 149 |
| 7.8    | Energy Recovery from High Pressure Streams          |      | 150 |
| 7.9    | Insulation                                          |      | 150 |
| 7.10   | Plant Layout                                        |      | 151 |
| 7.11   | Concluding Remarks                                  |      | 151 |

| **8**  | **Materials Recycling**                             |      | **155** |
| 8.1    | Introduction                                        |      | 155 |
| 8.2    | Recycling of Materials in Chemical Processes        |      | 155 |
|        | 8.2.1 | Economics of Recycling Process Streams               | 156 |
|        | 8.2.2 | Environmental Credits and Burdens of Recycling       | 156 |
| 8.3    | Closed Loop and Open Loop Recycling                 |      | 157 |
| 8.4    | On-Site and Off-Site Recycling                      |      | 159 |
|        | 8.4.1 | Examples of Off-Site Recycling                       | 159 |
| 8.5    | Producer and Consumer Waste                         |      | 159 |
| 8.6    | Hierarchical Approach to Materials Recycling        |      | 160 |
| 8.7    | Plastics Recycling                                  |      | 161 |
| 8.8    | Glass Recycling                                     |      | 163 |
| 8.9    | Recycling of Materials from Products                |      | 164 |
| 8.10   | Waste Treatment Option                              |      | 164 |
| 8.11   | Aqueous Effluent Treatment and Water Recycling      |      | 165 |
| 8.12   | Disposal of Wastes                                  |      | 167 |
|        | 8.12.1 | Landfill                                            | 167 |
|        | 8.12.2 | Incineration                                        | 168 |
| 8.13   | Concluding Remarks                                  |      | 169 |

**9 Waste Minimisation in Operations**     **171**
9.1   Non-Flow-Sheet Emissions from a Process Plant     171
9.2   Plant Start-Up     172
     9.2.1   Case Example — Starting Up a Sulphuric Acid Plant     172
9.3   Shut-Down of a Plant     173
9.4   Abnormal Operation     173
9.5   Plant Maintenance     174
9.6   Cleaning of Plant and Equipment     175
9.7   Fouling     175
9.8   Transport and Storage of Raw Materials and Products     176
     9.8.1   Storage Tanks     177
     9.8.2   Major Environmental Incidents Arising from Storage     178
9.9   Fugitive Emissions     179
9.10 Environmental Risks Resulting from Storm Water     180
9.11 Risks in Mining and Extraction of Materials     180
9.12 Concluding Remarks     181

**Problems: Part B**     **183**

PART C: EVALUATION

**10   Life Cycle Assessment**     **193**
10.1   Introduction     193
10.2   Product and Process Applications     194
10.3   Basic Steps in Life Cycle Assessment     196
10.4   Goal Definition     197
     10.4.1   Example of System Boundary Determination     198
10.5   Inventory Analysis     201
     10.5.1   Treatment of Utilities and Energy     202
     10.5.2   Allocation Procedures     202
10.6   Example of Inventory Data Estimation     203
10.7   Classification     205
     10.7.1   Further Discussion of Impact Categories     206
     10.7.2   Assignment and Weighting of Chemical Compounds     210
     10.7.3   Normalisation     215
10.8   Improvement Analysis     215
10.9   Some Challenges and Uncertainties in LCA     217
     10.9.1   Goal Definition     217
     10.9.2   Inventory Data     217

10.9.3  By-Products — Marketable or Waste?                 219
10.9.4  Impact Analysis                                    219
10.9.5  Resource Depletion                                 220
10.9.6  Normalisation                                      221
10.9.7  Valuation                                          222
10.10 Some Alternative or Supplementary Approaches to LCA   224
        10.10.1 EPS System — An Example of Evaluation Used
                with Inventory Data                         224
        10.10.2 Eco-Indicator                               225
10.11 LCA Software                                          225
10.12 Concluding Remarks                                    225

**11  Life Cycle Assessment Case Studies                    229**
11.1  Introduction                                          229
11.2  Life Cycle Inventories for Common Utilities           229
        11.2.1  Assumptions by Golonka and Burgess Regarding
                Utility Systems                              230
        11.2.2  Derived Inventory Data                       231
11.3  Inventory Data for Distinct Electricity Supply Systems 231
11.4  Hydrotreating of Diesel                               233
        11.4.1  Hydrotreating Process                        233
        11.4.2  System Boundary                              233
        11.4.3  Inventory Data                               235
        11.4.4  Impact Assessment                            239
        11.4.5  Environmental Burden versus Benefit Comparison 242
        11.4.6  Conclusions                                  243

**12  Safety Evaluation                                     247**
12.1  Introduction                                          247
12.2  Importance of Learning from Accidents, Dangerous
      Occurrences                                           249
12.3  Life Cycle Issues                                     251
12.4  Health, Safety, and the Environment                   252
12.5  Examples of Safety Incidents in the Process Industries
      Involving Environmental Damage                        253
12.6  Accident Prevention                                   255
        12.6.1  The HAZOP Approach                           255
12.7  Techniques for Investigating Probability of Major Incidents 256
12.8  Risk Assessment                                       257

12.9  Preventive Approaches                                           258
12.10 Transport and Storage of Chemicals                              259
12.11 Government Legislation                                          260
12.12 Concluding Remarks                                             261

13  **Assessment of Costs and Economics**                            **265**
    13.1  Introduction                                               265
    13.2  Investment Projects                                        266
    13.3  Capital Requirements and Sources                           267
    13.4  Fixed Capital Costs                                        268
          13.4.1  Approximate Estimates of Plant                     268
          13.4.2  Equipment Costs                                    270
          13.4.3  Contributions to Plant Costs                       271
          13.4.4  Estimating Fixed Capital Costs of Plants           271
    13.5  Working Capital Costs                                      273
    13.6  Operating Costs                                            274
          13.6.1  Simplified Cost Model                              274
          13.6.2  Classification of Production Costs                 275
          13.6.3  Estimation of Production Costs                     275
          13.6.4  Depreciation                                       276
          13.6.5  Capital Recovery                                   276
          13.6.6  Worked Example: Cost of Utilities Generation       277
          13.6.7  Environmental Management-Related Costs             278
          13.6.8  Cost Sheet Summary                                 278
    13.7  Revenue or Benefits Estimation                             280
    13.8  Engineering for a Movable Target                           281
    13.9  Profitability                                              282
    13.10 Case Example                                               284
    13.11 Economies of Scale                                         289
          13.11.1 Case Example in Scale Economies                    290
    13.12 Environmental Externalities                                291
          13.12.1 Estimating External Environmental Costs            292
          13.12.2 Emission Taxes and Emission Trading Schemes        293
    13.13 Life Cycle Costs of Projects                               293
    13.14 Conclusions                                                294

14  **Sustainability Assessment**                                    **297**
    14.1  Introduction                                               297
    14.2  Common Threads in Enviro-Economic Assessment               297

14.3   Cost Benefit Approach to Enviro-Economic
       Assessment                                                    299
14.4   Quantifying Benefits and Burdens                             300
14.5   Case Examples in Enviro-Economic Assessment                  300
       14.5.1  Case 1: Sulphuric Acid Manufacture                   300
       14.5.2  Case 2: Product Improvement through
               Hydrotreating of Diesel                              302
       14.5.3  Case 3: Power Generation from Fossil Fuels:
               CCGT-NG versus ST-Br Coal                            304
14.6   Environmental Effects of Scale of Production                 307
14.7   Sustainability Assessment and Sustainability Metrics         308
       14.7.1  Case Study on Sustainability of Electricity
               Generation                                           309
14.8   Perception and Assessment of Risk                            310
14.9   Scenario Analysis                                            313
14.10  Concluding Remarks                                           313

**Problems: Part C**                                                **315**

PART D: IMPLEMENTATION

15  **Planning for Sustainable Process Industries**                 **325**
    15.1   Introduction                                             325
    15.2   Forecasting                                              325
    15.3   Scenario Development                                     326
    15.4   Technology Innovation                                   327
           15.4.1  Intensification                                  327
           15.4.2  Technology Diffusion                             328
           15.4.3  Technology Evolution                             328
           15.4.4  Rates of Change                                  329
    15.5   Transition to Renewable Feedstocks                       329
    15.6   Site Selection for Process Plants                        331
    15.7   Integration of Process Plants and Process Industries     333
    15.8   Distributed Manufacture                                  334
           15.8.1  Case of Aqueous Sodium Cyanide Production at
                   the Point of Use                                 335
    15.9   Government Legislation                                   340
    15.10  Stakeholder Engagement                                   341
    15.11  Lifestyle Implications                                   341

**16 Process Design and Project Development**      **345**
16.1   Introduction      345
16.2   The Design Process      345
16.3   Process Flow Sheet Development      349
     16.3.1   Defining the Need      349
     16.3.2   Creating Plausible Solutions      349
     16.3.3   Screening of Alternatives      350
     16.3.4   Further Evaluation of Selected Options      350
     16.3.5   Optimisation and Scrutiny of Final Solution      350
16.4   Criteria for Process Flow Sheet Evaluation      351
     16.4.1   Technical Feasibility      351
     16.4.2   Capital Cost      351
     16.4.3   Operating Costs      351
     16.4.4   Safety      352
     16.4.5   Environmental      352
     16.4.6   Sustainability      352
     16.4.7   Reliability      352
     16.4.8   Operability      353
16.5   Process Flow Sheet Documentation      353
16.6   Piping and Instrumentation Diagram      354
16.7   Project Development      354
16.8   Acceptability Criteria for Projects      356
     16.8.1   Technical Requirements      356
     16.8.2   Economic Viability      357
     16.8.3   Safety      359
     16.8.4   Environmental      360
     16.8.5   Sustainability      361
16.9   Integrating Criteria Assessments      362
16.10   Concluding Remarks      364

**17 Operations Management**      **369**
17.1   Operational Phase of a Process Plant      369
17.2   Plant Maintenance      370
17.3   Environment Management Systems      372
     17.3.1   Commitment and Policy      372
     17.3.2   Planning      372
     17.3.3   Implementation      372
     17.3.4   Measurement and Evaluation      374
     17.3.5   Review and Continuous Improvement      374

17.4   Environment Improvement Plan                             374
    17.4.1   Examples of EIPs                                  375
17.5   Responsible Care                                        375
17.6   Environmental Performance Monitoring                    376
17.7   Emergency Response Planning                             378
17.8   Sustainability Reporting                                378
17.9   Emissions Reporting                                     379
17.10 Concluding Remarks                                       380

**Problems: Part D**                                           **383**

*Index of Topics*                                              391
*Index of Cases*                                               397
*Index of Set Problems*                                        399

# Acknowledgements

A wide range of companies and individuals have assisted in various ways to enable me to write this book. My thanks are extended to

- Pan Stanford Publishing for initially encouraging me to write the book and for subsequently publishing it, and for the editorial work of Sarabjeet Garcha and his staff
- the Chemical Engineering department at Monash University for supporting my endeavour in writing the work
- Ms. Amanda de Ruyter for her constructive reading and review of the book in its draft form
- Dr. Andrew Hoadley for his contributions to the case study in Chapter 15
- Dr. I-Kwang Chang, Dr. John May, and Pan Stanford for their assistance with diagrams
- Dr. David Kearns, Dr. Andrew Hoadley, Ms. Katie Brown, Ms. Manjusha Thorpe, Dr. Kurt Golonka, Ms. Kerriden Pugh, and Ms. Amanda de Ruyter for their contributions in the reading and editing of proofs for various chapters of the book.

The book builds on the work of both my postgraduate and undergraduate students over the years, and they collectively contributed to case study material. In this context I particularly thank Dr. Kurt Golonka, Ms. Amanda de Ruyter, and Dr. John May.

A vast number of industrial and academic colleagues have assisted me, over the last twenty years in particular, in gaining an improved understanding of the implications of sustainability, cleaner production, and industrial ecology for our process industries. The openness and willingness of these colleagues to assist is a positive sign, and a basis for optimism that the ideals of sustainability will prevail in our profession and our process industries.

Finally, I thank my wife, Elaine, for her great patience during the writing and publishing of the book.

**David Brennan**

# Preface

At the time of writing this book, 50 years have passed since I commenced undergraduate studies in chemical engineering. The practice and evaluation of process engineering has changed radically since then, underpinned by technological change. There has always been emphasis on technical rigour and scientific foundation. Economics has always been a dominant criterion for project approval. Health and safety, both occupational and public, have always been key issues for plants and products. A marked change, however, has been the growing awareness of the environmental impacts of process plants and products, and approaches to assess and minimise these impacts. We have slowly learned that we cannot consume resources or emit wastes without risking damage to the quality of air, water, land, and living species, including humans. This awareness has progressed through various experiences and perceptions of environmental damage forms such as toxicity, acidification, photochemical smog, stratospheric ozone depletion, and global warming.

The process industries have a major environmental impact arising from their large scale of operation. Large quantities of raw materials and energy are consumed in the processing and transport of materials. A wide range of chemicals are emitted to the environment through process streams and indirectly from generation of utilities used in process plants. Each of the distinct stages in a product's life cycle, from extraction of raw materials to product use and disposal, contributes to the environmental impact of the product.

The term 'green process engineering' has been adopted to express the transformation in design and operating practices in the process industries needed to minimise environmental impact from a product's life cycle. The term 'sustainable process engineering' has been adopted to widen this concept to include economic and social impacts. A major characteristic of this transformation has been the widening of system boundaries beyond a process plant to incorporate the entire product life cycle, support systems providing utilities and treating effluents, and systems exchanging materials and energy streams. The terms 'meso scale' (plants integrated on a site) and 'macro scale' (plants integrated across different industries and sites) have been used to express the expanded horizon beyond the 'micro scale' or single plant.

A further characteristic of sustainable process engineering has been the broadening of evaluation criteria to include technical, economic, environmental, safety, social, and sustainability assessments, and the integration of these assessments. Such assessments are made progressively in a project's development, approval, and execution, but also in process engineering decisions.

Sustainable process engineering is directed towards existing processes, plants, and products through new technology development, design for change, and improved operations. Technology innovations are essential for progress to more sustainable products and processes, but must be supported by enabling skills in project participation and management, design, and operation.

Sustainable process engineering is presented in this book as four parts:

- fundamental concepts of sustainability, cleaner production, and industrial ecology with their characteristics and implications
- strategies for identifying and minimising waste in process plants, particularly reaction, separation, and utility systems and through energy conservation and materials recycling
- assessment methods for environmental, safety, economic, and sustainability criteria, and integration of these assessments
- procedures needed to implement change through planning, design, project development, and operations.

It is assumed that the reader of this book has fundamental knowledge in science and engineering, including mass and energy balances, thermodynamics, fluid mechanics, and heat transfer. It is further assumed that the reader will be acquiring knowledge and skills in chemical engineering beyond this book and will integrate the thinking and techniques in this book with that learning. This book is not intended to replace existing texts in fundamentals or design, but to supplement them. Hopefully the book will also encourage the reader to develop interests beyond chemical engineering in environmental science, environmental economics, and regulatory practices, which play a complementary role in achieving more sustainable processes.

Some cases and problems are provided to assist skills development in problem and systems definition, process development and flow-sheeting, systems documentation, and evaluation. There is opportunity to build on the principles, cases, and problems provided in the book to make more rigorous or detailed analysis, if desired. References have been provided to encourage supplementary reading.

Although process engineering is distinguishable by its quantitative approach, sustainable process engineering also involves qualitative concepts. It is important for professional engineers to have skills in arguing the case for sustainable options. This should be reflected in process engineering education and opportunity provided in student project work to develop this facility.

**David Brennan**
August 2012

# PART A

# CONCEPTS

# Introduction to Part A

In the first part of this book, we explore three key concepts of sustainable process engineering, those of *sustainability*, *cleaner production*, and *industrial ecology*. These are by no means the *only* concepts related to sustainable process engineering. Indeed an extensive suite of concepts and associated terms have evolved, many of which are related to environmental science and engineering, as well as to process engineering. This evolution of concepts has been strongly influenced by a growing awareness of environmental problems, their anthropogenic causes, and remedial actions. Some of these environmental problems include

- poor air quality and its impact on human health, especially in cities with large concentrations of people, transport, and industries
- acidification of forests and water ways, often related to movement of acid gas emissions prior to acid rain precipitation
- depletion of stratospheric ozone due to emission of ozone depleting substances, leading to increased UV radiation damage to humans, plant life, and materials
- global warming with its threats of extreme weather patterns, rising sea levels, and damage to diverse biological species
- perceived shortages of resources such as fossil fuels, minerals, water, and fertile land.

Responses to environmental damage and its causes have come through professions, including science and engineering, both from individuals and institutions, through environmental bodies and through legislation by government at national and world levels. Some examples of these responses are listed below, and enable us to gain an appreciation of the evolution of human awareness of environmental problems.

In the early 1960s, Rachel Carson published a book titled *Silent Spring* drawing attention to carelessness in the use of chemicals (particularly insecticides) in the environment (Carson, 1962). In 1972, use of DDT as an insecticide was essentially banned from agricultural practice in the United States of America.

In 1972, the Club of Rome published their report 'The Limits to Growth' on the effects of population and economic growth on resource consumption and waste emissions, and the inability of the planet to support such growth (Meadows 1972).

In 1976, a working party of the Institution of Chemical Engineers published a report exploring likely areas of material and energy resource shortages, especially those impacting on the United Kingdom, and their economic, social, and political implications (Institution of Chemical Engineers, 1976).

Notable milestones for environmental legislation in the United States of America include:

1955 Clean Air act was initially passed, followed by amendments in 1963, 1966, 1970, 1977, 1990. A major achievement over this period has been establishment of ambient air quality standards for CO, Pb, $SO_2$, $NO_2$, $O_3$, and particulates.

1970 Occupational Safety and Health Act was established to limit worker exposure to harmful or toxic substances, and to establish communication of hazards to users of chemicals including use of Material Safety Data Sheets.

1971 United States Environmental Protection Agency (USEPA) was established.

1972 Clean Water Act was passed with legislative authority for water quality in rivers, lakes, estuaries and wetlands.

1976 Toxic Substances Control Act was enacted to allow government assessment of the risk from introduction of new chemicals into commerce. Key aspects included inventory establishment for existing chemicals and notification procedure for new chemicals.

1976 Resource Conservation and Recovery Act was introduced to regulate facilities which treat, store, or dispose of hazardous waste.

1980 Comprehensive Environmental Response Compensation and Liability Act (also termed 'Superfund') was set up to deal with closed and abandoned sites for hazardous waste. This act was amended in 1986 to ensure consideration of state standards in site clean-ups and to increase funding.

1986 Emergency Planning and Community Right to Know Act was passed, stipulating annual reporting requirements by owners or operators of facilities. Arising from this, a toxic release inventory was established for each toxic chemical and each facility. The inventory stipulated quantities stored on site, released to the environment, or transferred offsite for treatment or disposal, and the treatment and disposal methods used.

1990 Pollution Prevention Act was passed establishing pollution prevention as the primary pollution management strategy for the United States. The act provided for a hierarchy of approaches in waste management: source reduction, recycling, safe treatment, secure disposal.

Useful detail of many aspects of environmental legislation in the United States are provided in Appendix A of Allen and Shonnard (2002).

There are similar threads through legislation in other countries. Within the United Kingdom approximately 3000 people died in the great London smog in 1952, when extensive domestic coal burning accompanied by unusual weather patterns caused a major increase in concentrations of smoke and acid gases in the atmosphere. Subsequent clean air legislation in 1956 and 1968 in the United Kingdom made the use of smokeless fuel obligatory (Darton, 2003).

An important development in environmental legislation in the United Kingdom was the UK Environment Protection Act 1990. Part 1 of the Act established two regimes:

- Integrated Pollution Control for most potentially polluting or technologically complex processes
- Local Authority Air Pollution Control for other cases.

Integrated pollution control (IPC) emphasised the importance on minimising waste emitted to all media incorporating air, water, and land. When a process was likely to involve release to more than one medium, the BPEO (best practicable environmental option) was required through the use of best available techniques not entailing excessive cost (BATNEEC). Previous UK regulatory approaches had involved separate acts for air, water, and land pollution, and separate government authorities enforcing those acts.

Integrated pollution prevention and control (IPCC) has also been an important part of European legislation, initially incorporated in the Directive on IPCC in 1996, and subsequently modified in 2008 (Directive 2008/1/EC). European legislation committed to sustainable development has continued to evolve since the early 1970s (Europa, 2010).

Important milestones in Australia include the initiation of environmental legislation and regulatory authorities within state governments in the early

1970s and establishment of the National Pollutant Inventory (NPI) in 1997. The NPI was established by the Australian Government as a public internet database, providing information on industrial emissions and transfers of nominated substances.

At an *international* level, there have been important milestones reflecting global concern about environmental problems, and the need for action on a unified and world scale. These include:

1987 Brundtland Report (WCED, 1987) expressed consensus by members of the World Commission (21 countries) on issues of environment and development. Much emphasis was on ecologically sustainable development and preserving biodiversity.

1987 Montreal Protocol on Substances that deplete the Ozone Layer where 23 nations agreed to cut usage of key chlorofluorocarbons (CFCs) by 50% by 1999. This was followed by London Amendment (1990) where 93 nations agreed to phase out production of CFCs and most halons by 2000.

1994 United Nations Framework Convention on Climate Change which provided the groundwork for Kyoto Protocol.

1997 Kyoto Protocol drafted a legally binding document on greenhouse gas abatement.

2002 Johannesburg Summit reaffirmed sustainable development as central to the international agenda and agreed to fight poverty and protect the environment. Linkages between poverty, the environment, and use of natural resources were emphasised. Specific targets, many with time frameworks were agreed.

2009 United Nations Climate Change Conference, Copenhagen, Denmark while generally failing to fulfil global aspirations, agreed to limit further warming to two degrees Celsius, linked developed and developing countries in agreement to cut emissions, and agreed to develop an international monitor of emissions.

Professional societies of scientists and engineers have responded to environmental challenges in various ways including the enunciation of key principles of green chemistry and green engineering. In 1998, the *Royal Society of Chemistry* launched the Green Chemistry Network to promote the practice and progress of green chemistry (http://www.greenchemistrynetwork.org/).

*The Institution of Chemical Engineers* has long had a commitment to safety and environmental protection, as well as more recent commitment to sustainability (http://www.icheme.org/sustainability/). Sustainability was a central theme of the Melbourne Communique (2001), and 'The Sustainability

Metrics' (2002) were published for use in the process industries. In the IChemE roadmap for 21st century (2007), key policy themes were identified including sustainability, safety, environment, energy, food and water.

In 2003, engineers and scientists convened at Sandestin Resort in USA for the first conference on 'Green Engineering: Defining the Principles' (Abraham, 2006). A set of nine principles (often referred to as the *Sandestin Principles*) was agreed to

1. engineer processes and products holistically use systems analysis and integrate environmental impact assessment tools
2. conserve and improve natural ecosystems while protecting human health and well-being
3. use life cycle thinking in all engineering activities
4. ensure all material and energy inputs and outputs are as inherently safe and benign as possible
5. minimise depletion of natural resources
6. strive to prevent waste
7. develop and apply engineering solutions while cognizant of local geography, aspirations and cultures
8. create engineering solutions beyond current or dominant technologies; improve, innovate, and invent (technologies) to improve sustainability
9. actively engage communities and stakeholders in development of engineering solutions.

Mindful of these historic developments and initiatives, we now examine more closely the concepts of sustainability, cleaner production, and industrial ecology.

Chapter 1 deals with sustainability concepts in a global context, including the importance of equity, the 'triple bottom line', and extended time horizons. The relevance of sustainability to the process industries and challenges in achieving sustainable outcomes are also addressed.

Chapter 2 deals with the concept of cleaner production, focussing on waste minimisation in process plants, the hierarchy of waste management, product life cycles, system boundary definition, and the sources of process and utility waste.

Chapter 3 addresses industrial ecology and the potential for waste minimisation through industrial planning, integrated site complexes, and integrated processing of surplus material and energy streams across industry boundaries.

## References

Abraham, M. A. (ed.) (2006) *Sustainability Science and Engineering. Defining Principles*, Elsevier, Amsterdam.

Allen, D. T. and Shonnard, D. R. (2002) *Green Engineering. Environmentally Conscious Design of Chemical Processes*, Prentice Hall, Upper Saddle River, NJ.

Carson, R. (1962) *Silent Spring,* Houghton Mifflin, New York.

Darton, R. C. (2003) Scenarios and metrics as guides to a sustainable future. The case of energy supply, *Trans IChemE*, Vol. 81, Part B, 295–302.

Europa (2010) *Summaries of environmental legislation.* http://europa.eu/legislation_summaries/environment/.

Institution of Chemical Engineers (1976) *Materials and Energy Resources,* IChemE, Rugby.

Meadows, D. H., Meadows, D. L., Randers J., and Behrens III, W. W. (1972) *The Limits to Growth: A Report for the Club of Rome's project on the Predicament of Mankind*, Universe Books, New York.

WCED (World Commission on Environment and Development) (1987) in *Our Common Future* (ed. Brundtland, G. H.), Oxford University Press, Oxford.

# Chapter 1

# Sustainability Concepts

## 1.1 The Concept of Sustainable Development

The concept of sustainable development emerged in the 1980's within the environmental movement. In 1980, three globally based environmental organisations (United Nations Environment Programme [UNEP], World Wide Fund for Nature [WWF], and International Union for Conservation of Nature and Natural Resources [IUCNNR]) jointly published 'The World Conservation Strategy' aimed at

(a) maintaining essential ecological processes and life support systems
(b) preserving genetic diversity
(c) ensuring the sustainable utilisation of species and ecosystems.

The publication (UNEP/WWF/IUCCNF, 1980) reported that 'for development to be sustainable, it must take account of social and ecological factors, as well as economic ones; of the living and non-living resource base; and of the long-term as well as the short-term advantages and disadvantages of alternative actions'.

In 1983, the United Nations General Assembly convened The World Commission on Environment and Development to address growing concern 'about the accelerating deterioration of the human environment and natural resources and the consequences of that deterioration for economic and social development'. It was not until the World Commission's report, commonly referred to as the Brundtland Report (after the commission's chairman)

*Sustainable Process Engineering: Concepts, Strategies, Evaluation, and Implementation*
David Brennan
Copyright © 2013 Pan Stanford Publishing Pte. Ltd.
ISBN 978-981-4316-78-1 (Hardcover), 978-981-4364-22-5 (eBook)
www.panstanford.com

was published in 1987, that the concept of sustainable development became generally recognised. The Brundtland Report defined sustainable development as 'meeting the needs of the present without compromising the ability of future generations to meet their own needs'. Most appraisals of sustainability refer back to this definition. Enshrined within this definition is the concept of *intergenerational equity*.

The Brundtland Report (WCED, 1987) also emphasised the importance of *intra-generational equity*: 'The Earth is one but the world is not. We all depend on one biosphere for sustaining our lives. Yet each community, each country strives for survival and prosperity with little regard for its impacts on others. Some consume the Earth's resources at a rate that would leave little for future generations. Others, many more in number, consume far too little and live with the prospects of hunger, squalor, disease and early deaths.'

The first chapter (Perdan, 2004) of 'Sustainable Development in Practice' (Azapagic, Perdan, and Clift, 2004) is an excellent introduction to the origins of the sustainability concept. In this chapter, there are stark examples of drivers towards sustainability. Some of these examples are from a Global Environment Outlook Report by UNEP (2002):

- Around 2 billion ha of soil was classed as degraded by human activities. About 1/6th of this was either 'strongly or extremely degraded' beyond restoration.
- Around half of the world's rivers were seriously depleted and polluted.
- Around 24% of mammals and 12% of bird species were regarded as globally threatened.
- Depletion of the ozone layer which protects life from UV light had reached record levels.
- Concentrations of carbon dioxide were 367 ppm or 25% higher than in 1850.

Other examples of drivers towards sustainability in the same chapter are taken from a UNEP report in 2002 on Human Development:

- Some 80 countries, amounting to 40% of the world's population were suffering from serious water shortages by 1995
- Around 1 billion people still lack access to safe drinking water and 2.4 billion to good sanitation (mainly in Africa and Asia)
- 2.8 billion people live on less than $2/day
- Every year, 11 million children die of preventable causes derived from poor nutrition, sanitation, material health and education.

These and many other indicators and trends imply that development on the global scale is unsustainable.

Since the Brundtland Report, some key world events shaping sustainability thinking include

- Earth Summit held in 1992
- Johannesburg Summit held in 2002.

The Earth Summit in Rio de Janeiro agreed to a global plan of action, Agenda 21, for achieving more sustainable development and agreed to some 27 supporting principles. Two of these principles, Principles 15 and 16 are commonly cited.

### ● The Precautionary Principle (15)
*Where there are threats of serious or irreversible environmental damage, lack of full scientific certainty shall not be used as reason for postponing cost-effective measures to prevent environmental degradation.*

### ● The Polluter Pays Principle (16)
*National authorities should endeavour to promote the internalization of environmental costs and the use of economic instruments, taking into account the approach that the polluter should, in principle, bear the cost of pollution, with due regard to the public interest and without distorting international trade and investment.*

Other widely cited principles (Hammond, 2000) include

### ● The Subsidiarity Principle (or Participative Principle)
*Widespread and informal public participation is an essential prerequisite for effective decision making. Political decisions should be taken at the lowest possible level, that is, as closely as possible to the citizen.*

### ● The Integration Principle
*Environmental requirements must be integrated into defining and implementing all aspects of policy making in government.*

At the Johannesburg Summit (2002) some tens of thousands of people participated, including heads of government and delegates from non-government organisations, businesses, scientific groups, trade unions, and youth, reflecting the ideals of broad participation and inclusiveness. Linkages between poverty, the environment, and use of natural resources were emphasised. Specific targets, many with time frameworks were agreed

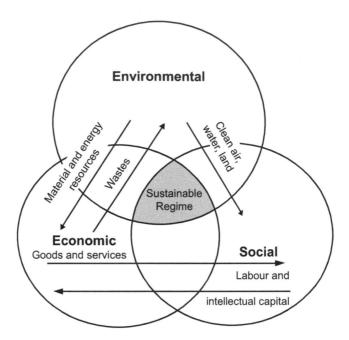

**Figure 1.1**  Environmental, economic, and social systems and their interactions. Brennan (2007): Reproduced with permission of John Wiley and Sons, Inc.

regarding poverty eradication, water and sanitation, supply and use of energy, use of chemicals, and management of natural resources.

While there is lack of unanimity about the exact meaning of sustainable development and its implications, most agree that it implies satisfying social, environmental, and economic goals. The term *triple bottom line* has been widely adopted by government and business in the context of meeting these goals. The characteristics of triple bottom line (Suggett and Goodsir, 2002) include the targeting, monitoring, and measurement of economic, environmental, and social performance, acceptance of accountability and transparency, and engagement of stakeholders.

Figure 1.1 shows a simplified view of *social, environmental,* and *economic* systems. These three systems are interconnected and interdependent. The economic system draws on materials and energy resources from the environment and on labour and intellectual capital from society. The economic system generates a diversity of goods and services for society's benefit, but at the expense of impacts on the environment, through both resource consumption and waste emissions. The environment provides clean air, water, and land, as well as the diversity of flora and fauna, to society; these provisions are essential to the development of the social system.

## 1.2 Sustainability in the Context of the Process Industries

In the context of the process industries, sustainability applies to whole industry sectors, products, process technologies, and individual process plants. In assessing performance achievement in sustainability, there is a need (Azapagic and Perdan, 2000) for suitable *sustainability indicators* to quantify performance and monitor progress in each of

- environmental
- economic
- social performances.

Work is proceeding to develop suitable scenarios and indicators for assessing sustainability. Many of the *environmental indicators* proposed are based on *life cycle assessment* methodology. They include, for example, indicators of

- resource depletion
- global warming potential
- ozone layer depletion
- photochemical smog
- human and eco toxicity.

*Economic indicators* proposed include

- value added
- contribution to gross domestic product
- capital expenditure, including that on environmental protection
- environmental liabilities
- ethical investments.

*Social indicators* proposed relate to ethics and human welfare, for example,

- stakeholder inclusion and participation
- international standards of conduct regarding business dealings, child labour
- income distribution
- satisfaction of social needs including work.

Stakeholder participation is consistent with the subsidiarity principle, and extends to a wide range of people encompassing product users, company shareholders, communities residing near plants, and other groups. Indicators of *safety performance* such as number of fatalities and lost time accidents should also be included under social indicators.

## 1.3  Some Temporal Characteristics of Sustainability

Sustainable development implies both *spatial* and *temporal* properties, because the criteria of sustainability must be met *locally* and *globally*, and for both *present* and *future* generations.

Sustainability implies a time horizon, which differs for a product, a process plant, a technology, or a whole industry. Technological transitions over time, in achieving improved products, processes, and plants, are necessary to ensure the sustainability of an industry. Some key aspects of time horizons and their implications are identified below.

### 1.3.1  *Time Horizons in Project Evaluation*

Since sustainability is concerned with intergenerational equity, time spans for evaluation of technologies, products, and processes should be extended beyond those which have traditionally been used within the process industries for project evaluation. Time spans for evaluating process plants have traditionally been 10–15 years, reflecting uncertainties in future trends of product and feedstock prices, feedstock availability, markets, and competing technology. *Longer term projections* involve exploring a wider range of possible scenarios for future developments.

### 1.3.2  *Time Horizons for Technology Development*

The development phase of a process technology from early research to commercial implementation is often lengthy, costly, and uncertain regarding technology definition, costs, market potential, scale development, and competing technology. This has important implications for implementing change where the current technology is environmentally deficient and requires major change.

### 1.3.3  *Time Dependence of Technology Improvement*

Learning, scale improvement, and cost reduction over time normally occur after a technology has been commercially adopted. The rate and extent of progress reflect the level of investment and competence in research, development and operational practice.

### 1.3.4 *Robustness to Technological, Economic, and Regulatory Change*

Since changes in markets, costs, prices, technologies, and company and government policies are inevitable over time, a desirable characteristic of a sustainable venture is robustness to change in these aspects over time. The extent of robustness can be explored by analysing sensitivity to a range of scenarios.

### 1.3.5 *Appraisal of Uncertainties (Technical, Business, and Environmental)*

Appraisal of uncertainties and their consequences (or *risk assessment*) is an essential component of sustainability assessment. This increases the time and cost of evaluation.

These temporal considerations run contrary to commercial opportunism of being able to capitalise promptly on perceived market opportunities. However, these are key aspects of sustainable development.

## 1.4 The Sustainable Project or Industry

To achieve a sustainable process industry project, the following aspects of the project are important and must be sustainable in their own right:

- products — function, cost, safety, environmental impact over life cycle
- feedstocks — availability, quality, cost, environmental impacts in extraction/refinement and impact on resource depletion
- energy sources — availability, quality, cost, environmental impact, impact on resource depletion
- labour, technical, and management personnel, adequately skilled
- technology — to enable design, construction, and operation of plant, and continuous improvement over project life
- capital availability and adequate return on investment.

Graedl (2002) identified four key items for sustainability in the chemical industry:

- chemical feedstocks
- energy for feedstock processing
- water for feedstock processing
- environment resilient to residual wastes emitted.

Water is identified because of the diverse demands for its uses (residential, commercial, agriculture), its variability in quality and availability, and its non-viability (in most cases) as a traded commodity.

Graedl recommended a strategy for a sustainable chemical industry:

1. Begin and maintain a transition from petrochemical to biochemical feedstocks.
2. Develop a strategy for limiting water use to a reasonable allocation of the locally available supply.
3. Begin and maintain a transition from fossil fuel and/or biomass energy to more sustainable energy.
4. Establish a program designed to achieve near zero discharges to the environment.

Graedl suggested a target time for sustainability of two generations (approximately 50 years), based on a report by the Board of Sustainable Development of the US National Research Council. He proposed time frameworks for transitions necessary to achieve sustainability. Target dates included

- 2035 for 50% contribution from petrochemical and 50% from biotechnological feedstocks for organic chemicals production
- 2040 for 50% contribution from fossil fuel and 50% from non-fossil fuel or essentially sustainable energy sources
- 2050 for close approach to zero discharges from processes to the environment.

## 1.5 Conflicts in Achieving Sustainability Objectives

We have seen that sustainability involves achieving standards in economic, environmental, and social criteria, and in intergenerational and intragenerational equity. We have focussed on the arguments and needs for more sustainable processes, plants, and industries. But we must acknowledge conflicting goals in the diverse spectrum of businesses, industries, and national and international economies which make progress towards sustainability difficult.

In the arguments for a carbon emissions trading scheme in Australia, we have seen the vested interests in lobbying from low carbon and carbon-intensive industries, from differing political ideologies and interest groups, as well as from professed climate change science 'acceptors' and 'sceptics' in government.

At the 2009 United Nations Climate Change Conference in Copenhagen we saw the different positions adopted by richer developed countries typified by Europe and United States of America, emerging economic powers such as China and India, countries with poorer economies, and countries vulnerable to sea level rise and effects of severe climate. These different interests and needs made the achievement of policy agreement on climate change difficult.

In an editorial for a special topic issue on Sustainable Development and Technology, in the journal 'Process Safety and Environmental Protection', Sharratt (2003) identified a number of obstacles to achieving sustainable development in practice. The editor further commented, 'The chemical engineering community has not transformed en bloc to address or even embrace sustainability'. In 2004, as part of its third Assembly, the IChemE debated the motion 'the commitment of the chemical engineering profession to sustainability pays no more than lip service to an ideal and has little impact on industry and society'. The motion was narrowly carried after arguments for and against it were conducted (IChemE Assembly, 2004).

It is clear that a high level of expertise, resolve, and cooperation between different groups of people in nations, governments, industries, and professions is needed to overcome the challenges and obstacles in achieving sustainable outcomes. There is need for an ideological shift towards individuals and groups owning the commitment to explore and implement sustainable practices. This extends to chemical engineering in terms of technologies, processes, plants, products, and related industries as well as professional practices. Chemical engineers also have a responsibility in engaging and educating the wider community in terms of sustainable lifestyle, product, and technology choices.

## References

Azapagic, A. and Perdan, S. (2000) Indicators of sustainable development for industry: a general framework, *Transactions IChemE*, Vol. 78, Part B, July, 243–261.

Azapagic, A., Perdan, S., and Clift, R. (eds) (2004) *Sustainable Development in Practice,* Wiley, Chichester, UK.

Brennan, D. (2007) Life cycle evaluation of chemical processing plants (Chapter 3), in *Handbook of Environmentally Conscious Materials and Chemical Processing* (ed. Kutz, M.), John Wiley and Sons, Hoboken, NJ, pp. 59–88.

Graedl, T. E. (2002) Green chemistry and sustainable development (Chapter 4), in *Handbook of Green Chemistry and Technology* (ed. Clark, J., and Macquarrie, D.), Blackwell Publishing, Oxford, pp. 56–61.

Hammond, G. P. (2000) Energy, environment and sustainable development: a UK perspective. *Trans IChemE*, Vol. 78, Part B, July, 304–323.

IChemE Assembly (2004), Assembly tackles sustainability, *The Chemical Engineer*, June 2004, p. 22.

Perdan, S. (2004) Introduction to sustainable development (Chapter 1), in *Sustainable Development in Practice* (ed. Azapagic, A., Perdan, S, and Clift, R.), John Wiley and Sons, Chichester, UK, pp. 3–28.

Sharratt, P. (2003) Editorial. Special Topic Issue – sustainable development and technology, *Transactions IChemE*, Vol. 81, Part B, September, 281–282.

Suggett, D. and Goodsir, B. (2002) *Triple Bottom Line Measurement and Reporting in Australia*, The Allen Consulting Group, Melbourne. Published by Commonwealth of Australia.

UNEP/WWF/IUCNNF (1980) *World Conservation Strategy. International Union for Conservation of Nature and Natural Resources*, www.iucn.org.

UNEP (2002) *Global Environment Outlook, United Nations Environment Program*, Earthscan Publications, London.

WCED (World Commission on Environment and Development) (1987) in *Our Common Future* (ed. Brundtland, G. H.), Oxford University Press, New York.

# Chapter 2

# Cleaner Production

## 2.1 Introduction

Achieving higher levels of sustainability in industry demands focus across a range of activities. These include developing new technologies, improving existing technologies, designing new plants, revamping existing plants, and improved operation of existing plants. Cleaner production is an important concept for achieving more sustainable products and processes in these contexts. Cleaner production emerged as a concept in the late 1980s and has been incorporated into both environmental legislation and industrial planning as a key pathway towards more sustainable performance.

## 2.2 The Concept of 'Cleaner Production'

A definition proposed by United Nations Environment Program (1992) at its first Asia Pacific Conference on Cleaner Production identified the following characteristics of cleaner production:

- application of an integrated preventive environmental strategy to processes and products to increase efficiency and reduce risks to humans and the environment
- for processes, reduction in consumption of raw materials and in the quantity and toxicity of wastes before they leave a process
- for products, reduction in impacts along the entire product life cycle, from raw material extraction to product disposal

*Sustainable Process Engineering: Concepts, Strategies, Evaluation, and Implementation*
David Brennan
Copyright © 2013 Pan Stanford Publishing Pte. Ltd.
ISBN 978-981-4316-78-1 (Hardcover), 978-981-4364-22-5 (eBook)
www.panstanford.com

- application of know-how, improvement in technology, and change in both management attitudes and operating practices.

This definition included several key points.

*Integration* — Cleaner production initiatives must be integrated

- with business objectives, duly recognising the needs of industry employees and the wider community
- with technical, safety, and operational considerations of processes and plants
- over the complete spectrum of industry activity from planning, design, construction, through to operation and management.

*Prevention* — Cleaner production emphasises *elimination* or *reduction* of waste *at source*. The term *pollution prevention* is also widely used in this context, particularly in the United States of America. This approach is vastly preferred to *end-of pipe* treatment of a terminal waste stream from a process, because it avoids the environmental impacts and costs associated with separating the waste and subsequent treatment of the waste. The concept of *integrated pollution prevention and control*, widely used in Europe, implies minimisation of damage from waste emission to the entire environment, through air, water, and land media.

*Product Life Cycle* — Cleaner production is applied to the entire life cycle of a product from extraction of basic raw materials, through material processing stages, product assembly, packaging and distribution, use, and final disposal.

*Global Applicability* — Although there are differences in priorities and approaches, there is *international acceptance* of the need for cleaner production, and implementation of programs by government and industry to achieve cleaner production outcomes.

The principles of cleaner production apply to

- processes and products of industry *including, but not limited to* process industries
- all industrial, commercial and domestic activity including, for example, hospitals, building and construction, transport.

## 2.3 The Product Life Cycle

Figure 2.1 shows a simplified representation of the key steps for a generalised product life cycle. Additional steps may occur in materials transportation and in materials recycling.

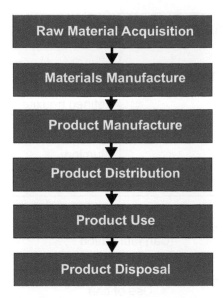

**Figure 2.1**   Key steps and sequence of the life cycle of a product. Brennan (2007): Reproduced with permission of John Wiley and Sons, Inc.

Various terms are often used for segments of the product life cycle. Thus

- *cradle to grave* encompasses the entire life cycle
- *cradle to gate* encompasses raw materials extraction through a processing chain to a product leaving the factory gate
- *gate to gate* encompasses a manufacturing process at a particular site.

Figure 2.2 shows a simplified diagram of the specific example of the life cycle of an aluminium beverage can. The scope of the life cycle depicted in Fig. 2.2 is 'cradle to grave'. The scope of the life cycle from mining bauxite ore to production of aluminium metal at a smelter is 'cradle to gate'. The scope of the life cycle within an alumina refinery where bauxite is converted to alumina is termed 'gate to gate'.

Mined bauxite contains impurities which must be separated from alumina in an alumina refinery. Bauxite from Weipa in Queensland Australia, for example, contains approximately 55% $Al_2O_3$, 12% $Fe_2O_3$, 5% $SiO_2$, 3% $TiO_2$, and 25% $H_2O$. Mineral impurities are separated from alumina in the refinery using a caustic soda digestion process followed by separation of insoluble impurities, and precipitation and drying of alumina. The process involves significant consumption of energy and emission of gaseous, liquid and solid

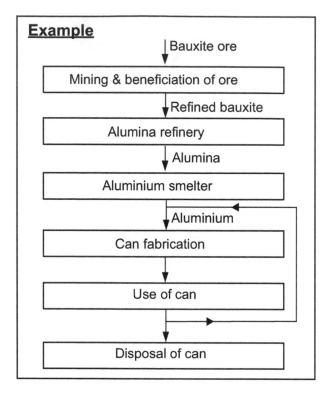

**Figure 2.2**  Product life cycle for an aluminium beverage can. Brennan (2007): Reproduced with permission of John Wiley and Sons, Inc.

wastes; a major waste is alkaline 'red mud' comprising the insoluble residues from ore digestion.

Dried alumina product from the alumina refinery is then converted to aluminium metal in a separate smelting process according to the equation

$$Al_2O_3 + 1.5\,C + electricity \rightarrow 2Al + 1.5\,CO_2$$

Aluminium smelters consume approximately 13 MWh of electricity per tonne of aluminium produced. If the electricity is generated from fossil fuel, major carbon dioxide emissions result in addition to the carbon dioxide emissions resulting from oxidation of the carbon anodes.

Note the materials recycling step in Fig. 2.2. Aluminium metal has the benefit of ease of recyclability, and its high strength to weight ratio has benefits in building, construction, and transport applications. The aluminium beverage can has environmental advantages in terms of light weight and ease of recycling after use, but some disadvantages in upstream impacts incurred through the mining of bauxite and its conversion through alumina to aluminium.

## 2.4 Hierarchy of Waste Management

A further key aspect of the cleaner production concept is the *hierarchy of waste management* illustrated in Fig. 2.3. This *hierarchy* may be designated as

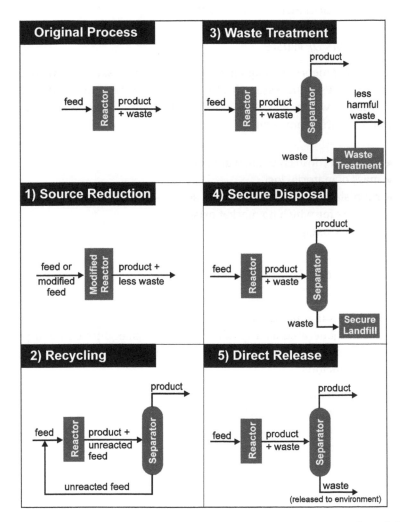

**Figure 2.3** Waste management hierarchy in chemical processes. Adapted from Allen and Rosselot (1997): Reproduced by permission of John Wiley and sons Inc.

follows with strategy 1 preferred to strategy 2, preferred in turn to strategy 3, and so on:

1. source reduction (eliminating or reducing waste at source)
2. recycling and reuse
3. waste treatment to render the waste less hazardous
4. secure disposal.

## 2.5 Concepts and Sources of Waste

### 2.5.1 *Concepts of Waste*

For chemical engineers, a major challenge is the minimisation of waste in the process industries. A necessary first step before waste minimisation is the identification of waste, including where and how it is formed. But what do we mean by waste? Some common concepts of waste, all of which are valid, include

- emission (or loss) to air, water, or land
- mass of material lost
- rubbish — lack of usefulness — scrap
- product for which no market exists
- energy wastage.

### 2.5.2 *Process and Utility Waste*

*Waste* occurs in both *process* and *utility systems*. The process typically comprises reaction, separation, heat exchange, and materials transport steps. Waste may be generated as a result of the process, equipment, and piping within the physical plant, or in product packaging and distribution. Process waste sources include

- feedstock impurities
- waste generated in reactors
- wastes generated in separation equipment
- energy required for the process
- extraneous materials such as catalysts, solvents, filter-aids, adsorbents
- packaging for raw materials and products.

Process waste streams may be emitted to the environment continuously or intermittently as

- purges
- vents and drains
- leaks and spillages
- accumulated deposits
- spent extraneous materials such as catalysts or adsorbents.

An important aspect of waste minimisation is to avoid merely transferring pollutant from one medium to another. For example, in treating an effluent gas stream containing sulphur dioxide, we might use a strategy of scrubbing with NaOH to produce $Na_2SO_3$, or with lime to produce $CaSO_4$, leading to aqueous or solid wastes but must address any resulting water or land pollution. Likewise, we might remove traces of mercury from natural gas using an adsorbent, but must also address the problem of regenerating or treating the loaded adsorbent, without causing damage through mercury emissions to the environment.

Waste may be generated during

- normal operation of plant
- abnormal operation, start up or shut down
- maintenance and cleaning of equipment
- transport and storage of process materials.

Waste is also generated in the *utility systems* which support the process systems, mainly through energy provision.

## 2.5.3 *Utility Waste and System Boundary Definition*

An important consideration in accounting for waste generation is the definition of *system boundaries*. Figure 2.4 shows some system boundary considerations in the supply of utilities to a process plant.

Even though utilities may be generated remotely from the process plant site, the environmental burdens from utility generation may be justifiably attributed to the utility user. Examples of remote generation of utilities depicted in Fig. 2.4 where waste is formed include

- electricity generation from coal fired power stations and transmission to the plant
- natural gas extraction, purification, compression, and pipeline supply to the plant
- mining, cleaning, and transport of coal to a power station.

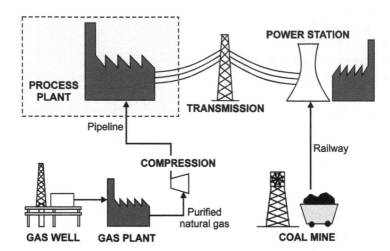

**Figure 2.4** System boundary considerations for waste from utility systems supplying process plants. Brennan (2007): Reproduced with permission of John Wiley and Sons, Inc.

Utility wastes on process plants are derived from provision of energy needs in the process. Main utility needs arise from

- *heating duties* including steam, hot oil, other heat transfer fluids, hot gases (usually produced from fuel combustion)
- *cooling duties* including cooling water, refrigeration
- *electricity* for drives on materials transport equipment (pumps, compressors, solids conveyors) and agitators, and in special cases for direct current supply in electrochemical processes.

Other utility needs include

- *nitrogen* for purging
- *compressed air*
- *water* for cleaning, fire fighting.

Utility waste can be minimised *both by*

- reducing consumption of utilities on process plants and
- improving the design of the utility systems themselves.

Utility wastes make a major contribution to wastes attributable to a given process. When *aggregated* over the whole life cycle of a *product*, utility wastes make a major contribution to the environmental impact of the product.

### 2.5.4 *Packaging*

System boundary definition is also important in relation to product packaging. A chemical plant may use packaging for its product, but in so doing contributes to any wastes emitted in the production, use, recycling or disposal of the packaging.

## 2.6 Impacts of Waste

Wastes differ in chemical composition, in phase (solid, liquid, or gas), and in terms of the medium to which they are released (air, water, or land). The impacts associated with release of wastes to the environment may be felt globally, regionally, or locally, for example:

- *global* — global warming, ozone layer depletion
- *regional* — acid rain, nutrification
- *local* — human toxicity.

Apart from this, different countries and regions may have distinct

- damage levels resulting from emissions, due to local differences in climate or topography
- environmental priorities, based on environmental, economic, or political considerations
- legislation dealing with waste emissions and cleaner production incentives.

Impacts of waste may be of short-term or long-term duration. An example of short-term duration is human toxicity effects from short-term release to atmosphere of a toxic gas. An example of long-term duration is land contaminated from solid waste disposal, affecting future generations.

## 2.7 Classification of Waste

Various approaches to waste classification have been adopted by different countries. In the United Kingdom, for example, waste has been classified as

- controlled waste (household, industrial, commercial)
- industrial waste ( from transport, utilities, telecommunications)
- special waste (process industry, hazardous wastes) including toxic, flammable, carcinogenic, corrosive, oxidising substances.

In the United States, some important classifications and related databases include

- criteria air pollutants ($SO_2$, $NO_x$, CO, $O_3$, Pb, particulate matter) where national ambient air quality standards (NAAQ) have been established. NAAQs are time-averaged concentrations which should not be exceeded in ambient air more than a specified number of times per year
- toxic chemicals reported under the toxic chemicals release inventory
- hazardous wastes — wastes defined as hazardous under the Resource Conservation and Recovery Act (RCRA)
- non-hazardous wastes — covered under Subtitle D of RCRA.

## 2.8 Driving Forces for Cleaner Production

Apart from achieving improved environmental quality, a number of driving forces encourage industry to adopt cleaner production practices. These include the following:

- *regulatory* through
  - ▶ international agreements
  - ▶ government at commonwealth, state, local levels
  - ▶ license to operate
  - ▶ need for environmental accreditation
- *customer perception*
- *company perception and policy*
- *employee perception*
- *community perception*
- *economic*, including
  - ▶ potential savings through reduced raw materials and energy use
  - ▶ reduced costs in generating and treating wastes
  - ▶ avoidance of future cost of remediation or waste treatment
  - ▶ minimization of potential liability arising from environmental damage.

## 2.9 Resistances to Introducing Cleaner Production

There are, however, potential obstacles to achieving cleaner production. Such obstacles can include

- resistance to change
- lack of skilled personnel to identify opportunities and develop solutions

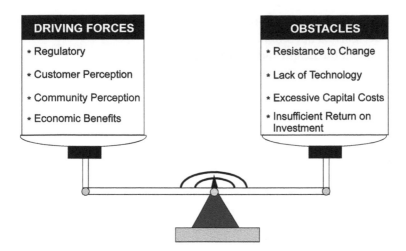

| DRIVING FORCES | OBSTACLES |
|---|---|
| * Regulatory | * Resistance to Change |
| * Customer Perception | * Lack of Technology |
| * Community Perception | * Excessive Capital Costs |
| * Economic Benefits | * Insufficient Return on Investment |

**Figure 2.5**   Some driving forces and obstacles in cleaner production initiatives.

- lack of appropriate technology
- time and cost of developing appropriate technology
- capital costs associated with implementing technical change
- insufficient return on capital investment from making technical change
- technical or business risks inherent in implementing change in technology, processes, plants, or operating procedures.

Figure 2.5 highlights the balance between driving forces for cleaner production and the obstacles to success, with some examples of contributing factors. Process engineering has a key role in creating a more favourable balance through providing improved technology and engineering practices.

## 2.10  Concluding Remarks

The characteristic objectives of cleaner production can be summarised as follows:

- Achieve reduction in the quantity of waste produced.
- Achieve reduction in the toxicity (or other damage property) of waste produced.
- Achieve reduction in consumption of raw materials, energy, and utilities.

- Address all stages of project life including research and development, design, and operation.
- Benefit the health and safety of workers.
- Apply to products (including their whole life cycle) as well as to processes.
- Extend to all industrial and commercial scale activities.

In order to achieve these objectives there are some key challenges:

- *Develop improved means of assessing or measuring 'cleanliness'* of processes or products.
- *Develop inherently cleaner processes*, for example, by changes in design, by using different or purer raw materials, or by substituting non-toxic materials for toxic materials.
- *Integrate* technical, safety, environmental, and economic objectives in process and product design; ensure that the benefits of cleaner production are identified and justified.
- *Specify appropriate system boundaries* for analysis, especially in relation to inclusions and exclusions.
- *Manage existing plants and systems* for cleaner performance and ensure continuous improvement.
- *Use better techniques* for design and revamping process plants.

Many of these objectives and challenges will be explored in later chapters and case studies.

## References

Allen, D. T. and Rosselot, K. S. (1997) *Pollution Prevention for Chemical Processes*, John Wiley and Sons, New York.

Brennan, D. (2007) Life cycle evaluation of chemical processing plants (Chapter 3), in *Handbook of Environmentally Conscious Materials and Chemical Processing* (ed. Kutz, M.), John Wiley and Sons, Hoboken, N. J., pp. 59–88.

United Nations Environment Program (UNEP) (1992) *Proceedings of Asia Pacific Cleaner Production Conference*, Melbourne.

# Chapter 3

# Industrial Ecology

## 3.1 The Basic Concept of Industrial Ecology

Ecology deals with living organisms — their habits, modes of life, and relations to their surroundings. Industrial ecology addresses the potential for better use of materials and energy in industrial systems. There is potential to improve efficiencies of industrial systems to approach those of natural ecosystems.

At present, industrial ecology is characterised predominantly by linear flows of materials and energy through inter-related production-consumption processes. There is need for a transition from an industry structure with predominantly linear flow of materials and energy to a structure with greater recycle of materials and energy, such as occurs with living organisms and *natural cycles*, such as those for carbon, oxygen, or nitrogen.

In industrial ecology, energy and material flows are part of individual production processes which interact with other industries and with consumption systems. Industrial systems have local, regional, and global dimensions and interact with natural ecosystems on these scales.

*Scales (or scopes) of influence* may be classified as microscale, mesoscale, or macroscale (Friedlander, 1994). An example of *microscale* activity is an individual process plant or section of a process plant. An example of *mesoscale* activity is a petrochemical complex or a petroleum refinery operating on a large site. An example of *macroscale* activity is the flow

*Sustainable Process Engineering: Concepts, Strategies, Evaluation, and Implementation*
David Brennan
Copyright © 2013 Pan Stanford Publishing Pte. Ltd.
ISBN 978-981-4316-78-1 (Hardcover), 978-981-4364-22-5 (eBook)
www.panstanford.com

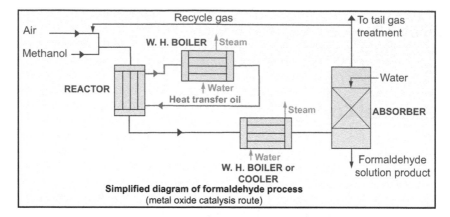

**Figure 3.1** Example of microscale activity. Simplified flow sheet of formaldehyde manufacture.

of industrial materials between diverse industry sectors such as power generation, chemicals manufacture, minerals processing, paper manufacture.

*Recovery and recycling of waste* are important elements of reducing both consumption of materials and energy, and emissions of waste. This applies not only within a single plant but also across plant, site, and industry boundaries.

Figure 3.1 shows an example of microscale activity. Formaldehyde is manufactured by the catalytic oxidation of methanol. The reaction occurs at elevated temperature; hot gases leaving the reactor are cooled in a waste heat boiler generating steam; formaldehyde is then absorbed in water in an absorber. Unreacted methanol is recycled to the reactor feed. Steam generated in the waste heat boiler can be used for driving pumps and compressors in the formaldehyde plant. Steam could also be exported to another industrial plant if this was feasible.

Figure 3.2 shows an example of mesoscale activity. Basic raw materials such as salt and naphtha are processed into intermediate chemicals such as ethylene and chlorine. These intermediate chemicals are then used as feedstocks for processing into a range of chemical products. All process plants are located on the same site. Surplus steam from some plants can be exported to other plants on the site which require steam for processing. Potential also exists for waste from one plant to be used as a raw material in a separate plant.

Figure 3.2 indicates the structure of the complex as it was in earlier years. Since then, changes have occurred in the structure of the complex and the size and technology of some individual plants due to changes in feedstock availability, product markets, and process technology. It is a feature

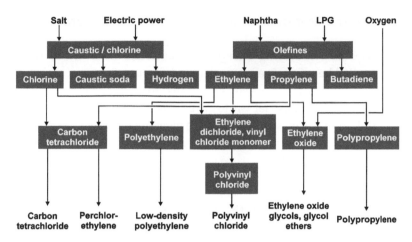

**Figure 3.2**   Example of mesoscale activity. Schematic outline of earlier operations at a petrochemical complex of ICI (Australia) at Botany, NSW.

of chemical processing complexes that they must adapt to such changes over time.

Figure 3.3 shows a generalised example of macroscale activities. Surplus material and energy streams from individual plants can be used on other plants but across multiple sites and industry sectors. This results in a net reduction of resources consumed for both process feedstocks and utilities and a net reduction in wastes emitted.

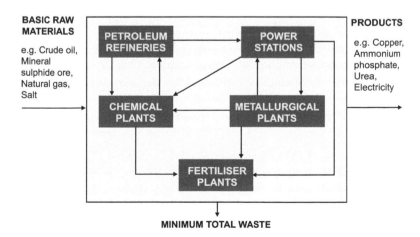

**Figure 3.3**   Example of macroscale activity. Materials and energy flows between plants in distinct industry sectors. Brennan (2007): Reproduced with permission of John Wiley and Sons, Inc.

## 3.2 Energy and Materials Recovery from Waste Streams

Opportunities for recovering energy and materials from waste streams provide the dual benefits of reducing demand on virgin sources of materials and energy, and reducing waste emissions to the environment. Some examples related to industrial wastes are provided by Allen (2004) while appraisals of domestic waste treatment are given by Azapagic, Duff, and Clift (2004) for the case of sewage water, and Kirkby and Azapagic (2004) for the case of municipal solid waste.

## 3.3 Resource Flow through the Economy

The flow of a particular resource through the industrial economy is key information for improving its industrial ecology. The flow of sulphur, for example, is a case of predominantly linear flow. This is partly related to its perceived low value in the economy, and relatively benign nature as a waste (as sulphur or sulphate, though not as sulphur dioxide). A much greater proportion of lead (70%) is recycled, due to its higher value and hazardous nature as a waste. Similarly a high proportion of platinum is recycled, due to its high economic value (Allen and Rosselot, 1997).

### 3.3.1 *Sulphur Flow in Australia*

The sulphur balance for Australia for the year ending June 1992 is shown as an example (Fig. 3.4). The balance indicates considerable sulphur loss to atmosphere as $SO_2$. Such balances can change markedly with time. The 1992 balance shows only minor imports of sulphur. This contrasts with large sulphur imports in 1980 and 2000. Much of the $SO_2$ released in 1992 is now captured to make sulphuric acid.

There is little recycling of sulphur-derived products. The bulk of sulphur is converted to sulphuric acid, which diffuses through the industrial economy. Much of sulphuric acid is used for making fertilisers. Other uses are for neutralising, gas drying, and alkylation within chemical and petroleum industries, and for lead acid batteries. Technology is available to reconcentrate dilute sulphuric acid (vacuum evaporation) and to regenerate spent alkylation acid (furnace and sulphuric acid plant). However, waste sulphuric acid is an unattractive prospect for recycling, because of its low value and relatively high cost of transport from point of generation to processing centres. It is also often contaminated.

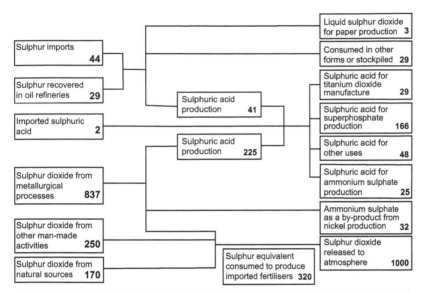

Note: liquid sulphur dioxide imports and production, potassium sulphate imports, elemental sulphur exports, fertiliser exports and losses to slags and matters are small and are not shown.

Australian sulphur balance for the financial year 1991–92. All values are in kilotonnes of sulphur equivalent.

**Figure 3.4** Example of a national commodity balance. Sulphur in Australia for financial year 1991–1992. All values are in kilotonnes of sulphur equivalent. Golonka and Brennan (1994): Reprinted with permission of AusIMM.

## 3.4 Transport and Storage of Raw Materials and Products

A number of environmental burdens and risks derive from the transport and storage of process industry cargoes. With all transport modes, there are impacts of fuel combustion and engine wear. For road and rail transport, there are noise considerations, and wear from friction on roads and rail supports. A major concern is the risk of spillage from the transport container, where safety risks also apply in the case of hazardous materials. Ship inventories are largest and damage potential is considerable. Examples of previous marine accidents include:

| Year | Location | Vessel | Loss |
|------|----------|--------|------|
| 1978 | Brittany | Amoco Cadiz | 230 kt crude oil |
| 1989 | Alaska | Exxon Valdez | 44 kt crude oil |
| 1993 | Shetland | Braer | 87 kt crude oil |
| 1994 | Baltic Sea | Estonia | 910 human lives |

### 3.4.1 *Marine Transport*

In the case of marine transport, potential leakage sources include

- leakage from ship of cargo or fuel resulting from storm, collision, or disrepair
- leakage during filling or emptying operations
- contamination of water (at destination) by ballast water (from port of origin)
- cargo leakage from storage vessels at despatch and arrival points.

### 3.4.2 *Road and Rail Transport*

Inventories of materials in road and rail transport are smaller than for ships, but many road and rail cargoes are potentially hazardous, for example, liquefied petroleum gas (LPG), liquid ammonia, vinyl chloride monomer. Transportation requires storage facilities at both despatch and arrival points with risks of containment loss associated with

- transfer of material into and out of storage
- loss of containment from storage vessels.

The magnitude of the potential hazard resulting from containment loss is related to the quantity stored. The magnitude of storage inventory in turn reflects transport frequency, and the associated supply/demand pattern for the material stored. Investigation of Australian chemical manufacturing sites indicated product storages of 15% annual production for stand-alone manufacture, and 2% annual production for site-integrated manufacture (Brennan, 1998). Safety and environmental risks of transport and storage, as well as economic benefits, are major driving forces for integrated manufacture on the one site.

## 3.5 Integrated Site Manufacture

*Integrated site manufacture* occurs where the product from plant A becomes raw material for plant B, and where plants A and B are located on the one site. *Integrated site manufacture* is to be distinguished from the case where plant A and plant B are on different sites requiring transportation of the intermediate chemical from plant A to plant B. As the number of integrated plants on a site is increased, potential benefits may also be expected to increase.

At the mesoscale and even more so at the macroscale, it is important to understand *potential markets* for surplus materials and energy streams. This requires a broad knowledge of process industries beyond the plant generating those surplus streams. It also implies an understanding of markets and end uses for intermediate products.

*Economic benefits* of integrated site manufacture arise from

- savings in fixed capital and working capital arising from reduced storage of intermediate products processed on site
- cheaper utility costs resulting from larger scale of utilities generation
- elimination of transport costs for intermediate products
- potential for surplus utility and process streams from one plant to be used on adjacent plants
- increased flexibility through use of shared buildings and personnel across the site.

*Safety and environmental benefits* of integrated site manufacture arise from

- transport of intermediates being confined within the site by pipeline
- avoiding risks to the public derived from off-site transport
- minimising risks in filling and emptying transport containers
- minimising risks derived from reduced storage of hazardous materials.

At the same time, there are potential constraints or disadvantages of site integrated manufacture which must be considered. For example,

- key raw materials or utilities may become available at a different location at lower prices
- integrated manufacture imposes high reliability requirements and operational constraints for individual plants because of their increased interdependence
- economic penalties can result from closure of a member plant on the site
- possible domino effects may occur in disaster propagation from a single incident
- cumulative effects of specific emissions (e.g., hydrocarbons) may cause an excessive environmental load at a given site.

## 3.6  Some Examples of Industrial Ecology Initiatives

### 3.6.1  *Case 1: Hydrogen Utilisation from Refineries*

In petroleum refineries, hydrogen is *produced* as a by-product from the catalytic reformer unit, used to increase the octane rating of naphtha produced from crude oil distillation. Hydrogen is *consumed* in refinery hydrotreaters, whose primary purpose is to remove sulphur from hydrocarbon streams.

There is often an imbalance between hydrogen supply and demand in refineries. Where there is surplus hydrogen, a common practice is to pipe the surplus hydrogen-rich hydrocarbon stream from the refinery to a nearby gas processing plant (Fig. 3.5). Here, the hydrogen is separated from hydrocarbons using pressure swing adsorption or cryogenic distillation. Hydrogen is then compressed, bottled and distributed to hydrogen markets, while hydrocarbons are piped back to the refinery.

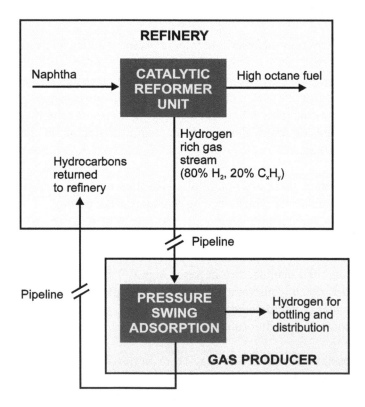

**Figure 3.5**   Recovery of surplus hydrogen from refineries.

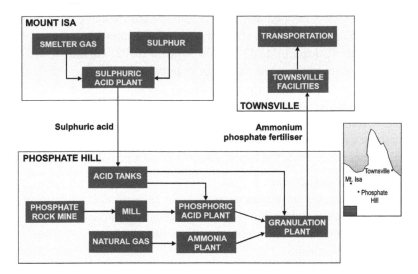

**Figure 3.6** Ammonium phosphate manufacture in Queensland. Brennan (2007): Reproduced with permission of John Wiley and Sons, Inc.

### 3.6.2 *Case 2: Fertiliser Complex, Queensland, Australia*

Phosphate rock from the North Queensland Duchess deposit is mined, beneficiated and reacted with sulphuric acid (Fig. 3.6). The sulphuric acid is produced from metallurgical smelter gases containing sulphur dioxide, at Mt. Isa. Acidulation of the rock produces phosphoric acid, which is separated from calcium sulphate by filtration. Key chemical equations are

$$SO_2 + 0.5O_2 + H_2O \rightarrow H_2SO_4$$
$$Ca_3(PO_4)_2 + 3H_2SO_4 \rightarrow 2H_3PO_4 + 3CaSO_4$$

Ammonia is produced by steam reforming of natural gas to make hydrogen which is then reacted with nitrogen. Phosphoric acid is reacted with ammonia to produce ammonium phosphate fertiliser, which is railed to Townsville on the Queensland coast. Key chemical equations are

$$CH_4 + 2H_2O \rightarrow 4H_2 + CO_2$$
$$3H_2 + N_2 \rightarrow 2NH_3$$
$$2\,NH_3 + H_3PO_4 \rightarrow (NH_4)_2HPO_4$$

The combined facility, with a capacity of 975 kt/yr high analysis ammonium phosphate fertilisers, commenced operation in 2000.

For many years prior to 2000, large quantities of $SO_2$ were emitted at Mt. Isa via a tall stack. Sulphuric acid was always a potential product, but no local market existed. Development of the phosphate rock deposit provided a sink for sulphuric acid. Imports of ammonium phosphate to Australia had increased from 50 ktpa in 1982 to 600 ktpa in 1992. Further growth in Australian demand has created a sizeable import replacement market, while opportunities also exist for exports. A natural gas pipeline from south-west Queensland to Mt. Isa created a source of methane for ammonia production. A rail link from Mt. Isa to Townsville and port facilities at Townsville provided infrastructure for materials transport to markets.

Several factors have combined to make such a project possible. These include

- *resource development*
  - ▶ natural gas extraction in south-west Queensland and natural gas pipeline construction (south-west Queens-land to Mt. Isa)
  - ▶ development of phosphate rock deposit
- *waste treatment*
  - ▶ $SO_2$ capture from smelter with greatly reduced emissions to produce sulphuric acid
- *exploitation of market* for ammonium phosphate
- opportunity to build on *transport infrastructure.*

### 3.6.3 *Case 3: Industrial Integration at Kalundborg, Denmark*

At Kalundborg in Denmark, four industrial facilities under different ownership and the local municipality participate in a system where wastes are profitably exchanged and utilised.

The participants are

- 1500 MW coal fired power station, owned by Asnaevaerket
- Denmark's largest refinery, owned by Statoil, with a capacity of 3 million tonnes of crude oil per annum
- Gyproc plaster board, producing 14 million $m^2$ of board per annum
- Novo Nordisk biotechnology, producing about 45% of world's insulin and 50% world's enzymes markets
- Kalundborg municipality, controlling the distribution of water, electricity and heating in the city area

Development at Kalundborg has been voluntary, but has occurred in close cooperation with government. Some key flows of energy and materials in 1995 are depicted in Fig. 3.7. In contrast to the Queensland fertiliser project which was planned as an integrated facility, integration at Kalundborg has evolved over time. The evolution has enabled major reductions in water consumption, in $CO_2$ emissions, and in solid wastes of fly ash and gypsum.

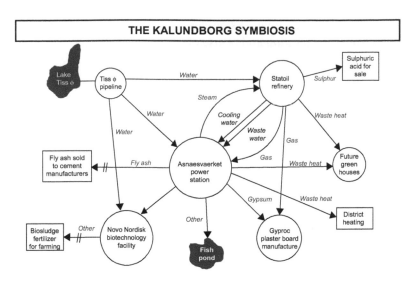

THE KALUNDBORG SYMBIOSIS

**Figure 3.7** Industrial integration at Kalundborg, Denmark. Grann (1997): Reprinted with permission.

### 3.6.4 *Case 4: Industrial Symbiosis at Kwinana, Western Australia*

Substantial and varied process industry development at Kwinana (south of Perth in Australia) has led to a number of regional synergies. Process plants at Kwinana (van Beers, Corder, Bossilkov, and van Berkel, 2007) include (as of 2007) the following:

- mineral processing plants including
  - ▶ 2000 kt/yr alumina refinery
  - ▶ 70 kt/yr nickel refinery
  - ▶ 105 kt/yr titanium dioxide plant
  - ▶ 800 kt/yr direct reduction iron making plant
- 850 kt lime and cement
- petroleum refinery processing 135,000 barrels/day of crude oil
- chemical processing plants producing
  - ▶ sodium cyanide
  - ▶ caustic soda and chlorine
  - ▶ ammonia and derived fertilisers

As of 2005, some 47 synergy projects were reported as being in place, including 32 by-product synergies and 15 utility synergies. Examples of by-product synergies reported include use of phosphogypsum from fertiliser production by the alumina refinery for assisting plant growth and soil stability in its residue areas. Examples of utility synergy include use electricity from cogeneration and re-use of treated waste water.

## 3.7 Concluding Remarks

Implications for sustainability and cleaner production extend beyond plant and site boundaries to industrial production and consumption systems. The flow of major industrial materials through the economy merits close scrutiny. Modelling tools are being developed and applied in order to assist this analysis. The recycling and exchange between industries of materials and energy demands industry cooperation, and a thorough evaluation of potential markets.

Transport of process materials incurs environmental impacts and risks of spillage. Opportunities for site integrated manufacture and for manufacture at the point of product use should be pursued, and have major implications for systematic industrial planning.

Industrial culture is evolving to explore and create an improved industrial ecology. Recent project initiatives provide evidence of this. However, there are potential economic and operational barriers to improved industrial ecology which should be recognised and addressed to ensure opportunities are realized. Some of these barriers derive from safety implications of transferring process and energy streams between different operating sites, and the possibility of intermittent stream flows with variations in flow rate and composition. Other potential derive from cost and profitability concerns.

## References

Allen, D. T. (2004) An industrial ecology: material flows and engineering design, in *Sustainable Development in Practice* (ed. Azapagic, A., Perdan, S, and Clift, R.), Wiley, Chichester, UK.

Allen, D. T. and Rosselot, K. S. (1997) *Pollution Prevention for Chemical Processes*, John Wiley and Sons, New York.

Azapagic, A., Duff, C., and Clift, R. (2004) Waste water management, in *Sustainable Development in Practice* (ed. Azapagic, A., Perdan, S, and Clift., R.), Wiley, New York.

Brennan, D. (1998) *Process Industry Economics: An International Perspective*. IChemE, Rugby .

Brennan, D. (2007) Life cycle evaluation of chemical processing plants (Chapter 3), in *Handbook of Environmentally Conscious Materials and Chemical Processing* (ed. Kutz, M.), Wiley, New York, pp. 59–88.

Friedlander, S. K. (1994) The two faces of technology: changing perspectives in design for environment, in *The Greening of Industrial Ecosystems* (ed. Allenby, B. R., and Richards, D. J.), National Academy Press, Washington DC, pp. 217–227.

Golonka, K. and Brennan, D. (1994) Supply and demand for sulphur in Australia. *Aus IMM Proc.*, No. 2, 69–76.

Grann, H. (1997) The industrial symbiosis at Kalundborg, Denmark, in *The Industrial Green Game. Implications for Environmental Design and Management* (ed. Richards, D. J.), National Academy Press, Washington DC, pp. 117–123.

Kirkby, N. and Azapagic, A. (2004) Municipal solid waste management, in *Sustainable Development in Practice* (ed. Azapagic, A., Perdan, S, and Clift, R.), Wiley, New York.

Van Beers, D., Corder, G., Bossilkov, A., and van Berkel, R. (2007) Industrial symbiosis in the Australian Mineral Industry. The cases of Kwinana and Gladstone, *J. Ind. Ecol.*, vol. 11, No. 1, 55–72.

# Problems

A number of problems are provided in each part of the book to illustrate the principles discussed in earlier chapters. The development of solutions to these problems builds on chemical engineering principles and systems, but in many cases there are a number of possible alternative solutions. Alternatives require evaluation and judgment, and a preferred solution requires reasoned argument. Information sources are suggested for some problems, but additional sources will generally be available. Websites suggested may change in terms of future access and detail, but will give an indication of potential useful sources derived from industry companies or bodies. The problems in Part A are intended to explore the concepts raised in the first three chapters. Some of the problem contexts reappear later in the book to explore principles discussed in later chapters.

## PROBLEMS: PART A

## 1. Sustainability of PVC Product

(a) Draw a life cycle diagram to show the entire product life cycle for polyvinyl chloride (PVC) pipe produced from basic raw materials of salt and ethane. The diagram should be a block diagram with one block for each distinct step of processing, usage, or disposal. Transport and distribution activities can be excluded. Identify on the diagram the main raw materials, intermediate products, co-products, and by-products from the various processing stages. Where possible, quantify the relationships between mass flows of raw materials, intermediate products and final product.

(b) Identify the important criteria by which PVC pipe should be judged in terms of being a sustainable product. Hence, postulate key advantages and disadvantages of PVC pipe in terms of being a sustainable product. Give your reasons.

(c) From an industrial ecology viewpoint, consider the sustainability merits of an integrated PVC polymer production facility (cradle to gate) based on ethane and salt raw materials and located at Karratha in northwest Australia, or at an alternative site of your choice. Identify the potential benefits, implications, and risks for such a scheme.

## Useful References

Ullmann (2002). *Encyclopaedia of Industrial Chemistry*, Wiley-VCH, Weinheim, Germany.

Websites for PVC manufacturers, for example:

European Vinyls Corporation: http://www.vinyl2010.org/ and http://www.vinylplus.eu/

Australian Vinyls: www.av.com.au

Vinnolit: http://www.vinnolit.de/vinnolit.nsf/id/EN_Home

## 2. Sustainability of Fuel Ethanol

Fuel ethanol is available in some countries as an alternative to petrol, either as a blended component (commonly 10% ethanol) or in some cases as a separate fuel. One source of raw material for fuel ethanol is sugar cane, which can provide either cane juice, a more concentrated syrup, or molasses by-product from sugar production.

(a) Draw a complete life cycle diagram for each of the following products:
- petrol derived from crude oil
- fuel ethanol derived from a sugar-based feedstock.

Life cycle diagrams should be simple block diagrams with one block for each distinct step of processing, transportation, usage, or disposal. Main inputs and outputs including co-products and by-products should be shown at each block.

For petrol, show major steps
- upstream of the refinery
- within the refinery
- downstream of the refinery, including use of petrol.

For fuel ethanol, show major steps upstream of sugar cane production, downstream of sugar cane production, within the ethanol 'distillery', and including use of fuel ethanol.

(b) Postulate *qualitatively* the major environmental impacts you would expect at each stage of the two product life cycles.

## Useful References

Institute of Petroleum (2002) *Modern Petroleum Technology,* Vol. 2: *Downstream* (A. G. Lucas, ed.), Wiley, Chichester, UK, Ch. 1, pp. 1–4, Ch. 9, pp. 83–85.

Moulin, J. A., Makkee, M., and Van Diepen, A. (2001) *Chemical Process Technology,* Wiley, Chichester, UK.

Ullmann (2002) *Encyclopaedia of Industrial Chemistry,* Wiley-VCH, Weinheim, Germany.

## 3. Hydrogen as an Alternative Car Fuel to Gasoline

With the prospect of increased price and reduced availability of crude oil and petrol worldwide, increasing attention is being given to alternative fuels for motor vehicles to supplement or replace gasoline (petrol). Petroleum-based diesel has been in use for some time, but LPG, compressed natural gas, fuel ethanol, and biodiesel have also been suggested and used. More recently, both electric cars and hydrogen fuelled cars have been suggested and some prototypes have been developed by particular car manufacturers.

(a) Draw a life cycle diagram for production, distribution and use of each of the following alternative automobile fuels:

  (i) gasoline (petrol) derived from crude oil

 (ii) hydrogen derived from the steam reforming of natural gas.

The diagrams in this case should be simple block diagrams with one block for each distinct step of processing, transportation, usage, or disposal. Main inputs and outputs including co-products and by-products should be shown at each block.

(b) (i) Identify four main sources of gasoline from within a petroleum refinery. For each source, identify and explain the main chemical or physical processing step.

  (ii) Briefly describe the steam reforming process for hydrogen manufacture from the steam reforming of natural gas. Include key chemical equations and indicate the equipment type at the main processing steps.

(c) Identify the main environmental impacts and their causes incurred in the *production* of

(i)  gasoline from crude oil

(ii)  hydrogen from the steam reforming of natural gas.

(d) Identify the emissions at the product usage step from using

(i)  gasoline

(ii)  hydrogen

as a car fuel. What environmental impacts result from these emissions?

(e) Discuss some important safety concerns arising from the distribution, storage, and use of each of the two alternative car fuels.

(f)  (i)  Nominate two alternative processes for the manufacture of hydrogen. Indicate one disadvantage of each process compared with the steam reforming of natural gas.

(ii)  Identify three existing markets for hydrogen apart from use as a fuel.

(g) Postulate the main environmental and sustainability issues arising from the use of the two alternative car fuels, hydrogen and gasoline.

## Useful References

Stefanakos, E. K., Goswami, Y., Srinivasan, S. S., and Wolan, J. T. W. (2007) Hydrogen energy in environmentally conscious energy production, (Ch. 7), in *Environmentally Conscious Alternative Energy Production* (Kutz, M., ed.), John Wiley and Sons, Hoboken, New Jersey.

Institute of Petroleum (2002) *Modern Petroleum Technology*, Vol. 2: *Downstream* (A. G. Lucas, ed.), Wiley, Chichester, UK, Chs. 1 and 9.

Ullmann (2002) *Encyclopaedia of Industrial Chemistry*, Wiley-VCH, Weinheim, Germany.

## 4. Industrial Ecology in Nitric Acid Manufacture

Ammonium nitrate is an important commercial chemical product used both as an explosive in the mining industry and a fertiliser.

(a) Identify the major stages in the life cycle of ammonium nitrate product, considered from 'cradle to gate' perspective. Assume that air, steam, and natural gas are the primary feedstocks. Express your answer using an appropriate block diagram.

(b) Postulate the major environmental impacts occurring at each stage of the ammonium nitrate product life cycle and their source.

(c) Identify the major downstream commercial uses for ammonia (other than ammonium nitrate) in the process industries.

(d) (i) Outline the industrial ecology implications for the design, site location, and integration (with other industrial plants or facilities) of a nitric acid manufacturing plant.

(ii) Recommend a site for a new nitric acid plant in your country or region which would offer benefits in terms of industrial ecology. Give reasons for your recommendation. Draw a block diagram showing the inter-relationship between individual plants, raw materials, intermediates, and products that you envisage.

## Useful References

Sinnott, R., and Towler, G. (2009) *Chemical Engineering Design*, Elsevier, Oxford, UK, Ch. 4.

Ullmann (2002) *Encyclopaedia of Industrial Chemistry*, Wiley-VCH, Weinheim, Germany.

## 5. Alternative Packaging for Beer

Beer is packaged in most countries in both glass bottles and aluminium cans. Both of these containers can be recycled after use, achieving environmental benefits. What are these benefits? Does either container have an advantage in the context of recycling? What other considerations might decide the preferred container for beer?

# PART B

# STRATEGIES

# Introduction to Part B

Having considered some key concepts shaping sustainable process engineering and their origins, we now explore strategies for waste minimisation in process plants. Considerable emphasis is placed on understanding how waste is generated at source. This understanding is an essential first step before developing strategies to minimise waste.

Consistent with the established 'hierarchy of process design' (often referred to as 'the onion diagram'), we begin by looking at *waste in reactors*. The reactor is often a key source of process waste influencing subsequent separation requirements, recycle of process streams, effluent quantities and compositions, and product quality. Product impurities in intermediate chemicals provide a further source of waste for downstream processing and downstream products. In Chapter 4, we consider the main sources of waste in chemical reactors; we explore how these sources occur in some selected chemical processes of industrial importance, and how these processes have evolved to minimise waste.

We then explore how waste is formed in some more common *separation processes* in the process industries and examine some specific case examples. Since energy is a major consideration in many reaction systems, as well as in separation processes and materials transport, we next examine the common *utility systems* which provide energy for process plants. Utilities are most widely used for heating, cooling and electrical duties. The generation of these utilities in turn involves fuel and water consumption and a range of waste emissions. We explore how energy and utilities are consumed in process plants and some of the ways in which energy consumption can be minimised.

Recycling, both of process streams within processes and of product materials in the wider economy, is an important strategy for minimising both resource consumption and waste emissions. Key aspects of recycling are explored, and related benefits and burdens, both economic and environmental, are identified.

We finally explore waste generation in the wider context of process industry operations, including departures from normal plant operating conditions, startup and shutdown of plants, leaks and spillages, cleaning and maintenance, as well as the storage and transport of process materials.

# Chapter 4

# Waste Minimisation in Reactors

## 4.1 Introduction

This chapter is structured in two parts. In the first part, we explore the principles governing waste generation and minimisation in reactors. In the second part, we explore these principles in the context of five commercial chemical processes.

Reactors involve chemical change and hence it is important to consider reaction chemistry, both in terms of what is intended and what actually occurs in practice. Reactors can be diverse and complex equipment, handling chemicals of widely different chemical and physical properties, operating at widely different conditions of temperature, pressure, and reactant/product concentrations, involving different phases (solid, liquid, and gaseous), having different geometries, and incorporating a wide range of distinct physical configurations and ancillary equipment. Nevertheless we can make some generalisations about sources of waste in reactors derived from

- stoichiometric generation of unsaleable co-products
- feedstock impurities
- imperfect mixing
- secondary reactions leading to undesired side products
- incomplete conversion of feedstock to products
- reversible reactions
- agent materials
- deterioration of catalysts during operation.

*Sustainable Process Engineering: Concepts, Strategies, Evaluation, and Implementation*
David Brennan
Copyright © 2013 Pan Stanford Publishing Pte. Ltd.
ISBN 978-981-4316-78-1 (Hardcover), 978-981-4364-22-5 (eBook)
www.panstanford.com

These sources of waste can be identified in industrial reactors to differing extents, but often several of the sources occur in any one chemical process.

After discussing these waste sources and some common design strategies for avoiding or minimising their effects, we consider five separate chemical processes. In each case, we explore the main chemical reactions, the main sources of waste and their consequences, and how the technology of the processes has evolved to minimise the generation of waste. The chemical processes examined are

- chlorine and caustic soda manufacture from salt
- ethylene manufacture from hydrocarbons
- hydrogen cyanide manufacture from ammonia, methane, and air and its conversion to sodium cyanide
- sulphuric acid manufacture from sulphur and mineral sulphides
- polyvinyl chloride (PVC) manufacture by suspension polymerisation of vinyl chloride.

All five processes are of commercial significance in the chemical industry. Chlorine, caustic soda, ethylene, and sulphuric acid are large tonnage products used as intermediate chemicals for further processing to make a wide range of downstream products. PVC is a large tonnage polymer with diverse applications while sodium cyanide is a key chemical in gold extraction.

## 4.2 A Checklist for Reaction Systems and Reactors

In exploring industrial reactors and their potential for waste generation, it is important to consider both the reaction chemistry including its thermodynamics and kinetics, and also the reactor system with its engineering features. Key aspects of reactor chemistry and reactor design provide a useful checklist.

- **Chemical reactions including**
  - ▶ basic chemistry of intended reaction or reaction sequence
  - ▶ whether reactions are reversible or irreversible
  - ▶ whether reactions are exothermic or endothermic
  - ▶ whether reactions are catalytic or non-catalytic
  - ▶ temperature, pressure, and concentration conditions for reactions
  - ▶ whether unintended side reactions are occurring
  - ▶ degree of conversion of reactants to desired products.

- **Reactor design**
  - ▶ phases under which the reaction proceeds (gas, liquid, solid, multiphase)
  - ▶ batch or continuous operation
  - ▶ provisions for heating and/or cooling
  - ▶ approach to a CSTR (continuous stirred tank) or plug flow reactor
  - ▶ reactor equipment type, for example, furnace, packed bed, electrolytic cell, stirred vessel, platinum gauze.

Some examples of reactor types and their use in different processes include

- furnace for
  - ▶ steam reforming of natural gas for hydrogen production
  - ▶ steam cracking of ethane to produce ethylene
- packed bed reactor for
  - ▶ sulphuric acid production (vanadium pentoxide catalyst)
  - ▶ isomerisation reactor in petroleum refining (alternative catalyst options)
- electrolytic cell for
  - ▶ chlorine and caustic soda production from sodium chloride
  - ▶ conversion of alumina to aluminium in aluminium smelting
- stirred vessel for
  - ▶ polymerisation of vinyl chloride in PVC production
- platinum gauze reactors for
  - ▶ oxidation of ammonia in nitric acid production
  - ▶ oxidation of ammonia and methane in hydrogen cyanide production.

These reactor types are used in the five chemical processes later reviewed. Sketches of platinum gauze and furnace reactors are provided in Fig. 15.3.

## 4.3 Chemistry of Process Route

Cleaner (or 'greener') chemistry can be regarded as a subset of cleaner production. Some of its objectives which impinge on reactors include

- developing synthesis pathways which use alternative feedstocks or more selective chemistry to generate less waste
- finding alternative reaction conditions, for example, lower temperatures which promote increased selectivity

- finding new solvents, for example, water might be used to replace an organic solvent
- finding improved catalysts
- developing alternative chemical products which are inherently safer or less toxic, providing benefits in storage, transport, and use.

A simple measure of environmental effectiveness of reaction chemistry is the *Environmental factor* (or *E factor*), where E factor = Ratio of kg waste (or by-products) to kg desired product. Typical values of E factors for industry sectors have been quoted by Sheldon (1994). These values indicate the relative magnitudes of waste in different industry sectors in United States at that time.

| Industry | Annual tonnage of products | E factor |
|---|---|---|
| Oil refining | $10^6$–$10^8$ | ~0.1 |
| Bulk chemicals | $10^4$–$10^6$ | <1 to 5 |
| Fine chemicals | $10^2$–$10^4$ | 5–50 |
| Pharmaceuticals | $10^1$–$10^3$ | 25–100 |

The E factor identifies the magnitude of waste, but does not distinguish between *harmful* and relatively *benign* wastes. Sheldon proposed an *environmental quotient*, $EQ = E \times Q$, where $Q$ is a factor for environmental impact. Life Cycle Assessment (LCA) and other techniques have been developed for estimating quantitative environmental indices for processes and products. LCA is discussed in Chapters 10 and 11.

*Atom utilisation* is a measure of how effectively chemical atoms participate in a reaction to achieve the desired product. Atom utilisation can be calculated on the basis of the stoichiometric equation by dividing the molecular weight of the desired product by the sum of molecular weights of all substances produced. An example of calculating atom utilisation is the case of ethylene oxide manufacture.

The overall reaction for the now obsolete chlorohydrin route for manufacturing ethylene oxide may be represented as

$$C_2H_4 + Cl_2 + Ca(OH)_2 \rightarrow C_2H_4O + CaCl_2 + H_2O$$

Molecular weight $\qquad\qquad$ 44 $\qquad$ 111 $\qquad$ 18

Atom utilisation $44/(44 + 111 + 18) = 0.25 = 25\%$

We can also look at utilisation for a particular atom, for example, oxygen, where utilisation is $1/(1 + 1)$ or 50%; *hydrogen*, where utilisation is $4/(4 + 2)$ or 67%. Atom utilisation for the chlorohydrin route can be compared with that of 100% for the modern direct oxidation route of ethylene to ethylene oxide.

$$2C_2H_4 + O_2 \rightarrow 2C_2H_4O$$

### 4.3.1 *Conversion, Selectivity, and Yield*

Conversion and selectivity are common measures of chemical reactor performance, and have major impacts on downstream separation and recycle.

*Conversion* is the ratio

$$\frac{\text{reactant consumed in a reactor}}{\text{reactant fed to the reactor}}$$

Limited conversion of feedstock implies feedstock is present in the stream leaving the reactor. From this point, it requires separation from the desired product and preferably recycling back to the reactor feed.

*Selectivity* is defined as the ratio

$$\frac{\text{moles of desired product produced}}{\text{moles of reactant consumed}} \times \text{stoichiometric factor}$$

where stoichiometric factor = stoichiometric moles of reactant required per mole of product.

Limited selectivity implies side reactions forming unwanted products in the reactor effluent; these unwanted products must be separated from the desired product. *Secondary reaction* can occur in *parallel* or in *series* with the desired primary reaction, and at distinct reaction rates.

In parallel,    Reactant → Desired Product
                          ↘ *Waste*

In series,    Reactant → Desired Product
                   *Desired Product → Waste*

Yield is another common measure of reactor performance. Yield is the ratio

$$\frac{\text{desired product produced}}{\text{reactant fed to reactor}} \times \text{stoichiometric factor}$$

Yield can be based on once through reactor performance, or within a system involving recycled feedstock. Clear definition of the basis of yield is therefore essential.

### 4.3.2 *Co-Product and By-Product Utilisation*

Co-products and by-products can generate sales revenue if markets for these products can be established and sustained. The term co-product indicates a second product of comparable mass quantity or revenue value with the primary product. The term by-product indicates a second product of rather lower revenue value than the primary product. If their markets fail, however, the co-product or by-product becomes a source of waste, and both an economic and environmental burden. Two examples of stoichiometric generation of by-product waste in primary reactions are now given. In each example the mass of by-product produced is approximately twice the mass of the desired product.

**Example 1. Manufacture of phosphoric acid by acidulation of phosphate rock.**
Phosphoric acid, a key chemical intermediate in the manufacture of phosphate fertilisers, is manufactured by the acidulation of phosphate rock. Gypsum is a low value by-product of this reaction.

$$Ca_3(PO_4)_2 + 3H_2SO_4 \rightarrow 2H_3PO_4 + 3CaSO_4$$

While there are markets for gypsum, mining of naturally occurring gypsum is usually the preferred source. 'Phosphogypsum' (produced in phosphate rock acidulation) is subject to contamination from impurities derived from phosphate rock. In many operations, phosphogypsum is stockpiled, representing a major solid waste (http://www.epa.gov/rpdweb00/neshaps/subpartr/about.html). An important strategy is to improve the quality of by-product gypsum to enable marketability or safe disposal. Possible approaches include purifying the phosphate rock feed and improving the washing of gypsum during filtration.

**Example 2. Propylene oxide manufacture by the chlorohydrin process.**
This process employs chlorine to make propylene chlorohydrin as an intermediate, in a similar way to the chlorohydrin route for making ethylene oxide. Lime is added to liberate propylene oxide from the chlorohydrin. A large quantity of calcium chloride by-product results, for which there is usually no market. The overall reaction can be represented by

$$C_3H_6 + Cl_2 + Ca(OH)_2 \rightarrow C_3H_6O + CaCl_2 + H_2O$$

## 4.4 Impurities in Reactor Feedstocks

Feedstock impurities can contribute to waste in several ways:

(a) Impurities in the feed can be unreacted, contaminating process streams downstream of the reactor;
(b) Impurities in one reactant can react with other reactants to produce a new waste species;
(c) Feed impurities can poison a catalyst in a reactor, leading to lower conversions and increasing the extent of secondary reactions.

In each case feedstock impurities can result in an effluent waste stream or product contamination. In (c) catalyst life is shortened, with economic and environmental consequences (see Section 4.9).

Impurities are characteristically present in all minerals and fossil fuels, even where these materials have undergone refinement before supply to process plants as feedstocks. These impurities, if not removed, can be fed forward into downstream processing chains. Impurities may also be introduced through other sources in a chemical process which in turn become impurities in feedstocks to downstream products. In most cases, it is best to remove an impurity as early as possible in the processing sequence and before the reactor, although there are some cases where removal after reaction is easier.

Table 4.1 gives some examples of impurities in crude oil (feedstock to a petroleum refinery), phosphate rock (feedstock to a fertiliser complex), sodium chloride salt (raw material for a chlorine-caustic soda plant) and bauxite (mineral feed for aluminium production). In the case of sulphur in crude oil, sulphur is distributed through the products of crude oil distillation, being most concentrated in the least volatile products. Sulphur then becomes an impurity in the bottoms product from the distillation column, which is fed in turn to a catalytic cracking unit. If not removed, sulphur contaminates the gasoline, diesel, and fuel oil products from the catalytic cracking unit.

Table 4.1    Examples of impurities in different basic raw materials

| Crude oil | Phosphate rock | Sodium chloride | Bauxite |
|---|---|---|---|
| Salt | Fluoride | Calcium | Silica |
| Sulphur | Silica | Magnesium | Ferric oxide |
| Nitrogen | Alumina | Iron | Titanium dioxide |
| Metals | Ferric oxide | Sulphate | |
| | Magnesium oxide | | |
| | Sodium oxide | | |

## 4.5 Mixing of Reactants

Thorough mixing of reactants assists in achieving better conversion and selectivity. It is also important to avoid short circuiting of reactor feed from the reactor inlet to the reactor outlet: this is assisted by careful design of reactor internals and location of feed and product nozzles.

### 4.5.1 *Mixing of Gaseous Reactants*

Static mixing devices located upstream of reactors can achieve greater homogeneity in both concentration and temperature of reactants than can be achieved solely through pipeline mixing. Static mixers have been used for mixing gaseous reactants in processes to make the following products:

(a) Nitric acid (for ammonia and air mixing)
(b) Sulphuric acid (for air and sulphur dioxide mixing)
(c) Ethylene (for mixing hydrocarbons and steam)
(d) Formaldehyde (for mixing methanol and air)
(e) Hydrogen cyanide (for mixing ammonia, methane and air).

In the cases of nitric acid, formaldehyde and hydrogen cyanide, uniform composition control is also important to ensure that the reactant compositions are outside flammability limits.

### 4.5.2 *Mixing of Liquids*

When mixing miscible or immiscible liquids, and when dispersing solids or gases in liquids, mixing quality is again important to achieve homogeneity in concentration or temperature. Special care should be taken in agitated vessels with choice of impeller type and speed, geometry of vessel and impeller, design of internals, and size and location of vessel nozzles. Jet mixing can be a worthwhile alternative to impeller mixing. For an agitated vessel needing heating or cooling, internal coils or an external jacket are common. In these cases, the effect of mixing on the heat transfer rate must be considered. Mixing is discussed in greater detail in Chapter 7.

### 4.5.3 *Fluid Distribution in Packed Bed Reactors*

When introducing gas or liquid feeds to packed catalyst beds, even distribution of fluid across the bed is important. Careful attention should be paid to the design of inlet piping and the distribution of feed over the packing.

## 4.6  Minimising Secondary Reactions

Secondary reactions can be minimised by improving the quality of reactant mixing before and within the reactor. Secondary reactions at elevated temperatures can also be minimised by having short contact times for the reaction followed by rapid cooling of the reactor effluent. Rapid cooling is often achieved in waste heat boilers located close to the reactor, and in some cases physically integrated with the reactor. Steam generated in the waste heat boiler can be used within the process for driving machinery or process heating; alternatively, it may be possible to export the steam to an adjacent plant.

Temperature constraints on process gases may exist in waste heat boilers. For example, exit temperatures may need to be below upper limits where side reactions are occurring, but also above lower limits where corrosive acids such as sulphuric or nitric acid might condense.

Design features of waste heat boilers can be varied. For example,

(a)  water may boil in the shell, as with vertical tubular reactors
(b)  hot oil systems may be used for reactor heating and cooling, with heat recovered from the oil to raise steam in a separate waste heat boiler
(c)  special materials of construction may be needed to withstand attack by process fluids.

Examples of processes using waste heat boilers in conjunction with reactors include

| Process | Approximate reactor temperature |
|---|---|
| Ethylene from ethane | 900°C |
| Nitric acid from ammonia and air | 900°C |
| Hydrogen cyanide from methane and air | 1100°C |
| Formaldehyde from methanol and air | 350°C (metal oxide catalyst) |

Direct quench systems, involving water addition to the reactor effluent stream, are sometimes used as alternatives to waste heat boilers. They can be effective in achieving sudden temperature reductions but should be evaluated for effects on

(a)  the resulting water balance in the process
(b)  generation of aqueous effluents
(c)  penalties in reduced energy recovery.

## 4.7 Recycle of Unreacted Feed from Reactor Outlet

For cases of limited reactant conversion, the recycling of unreacted feed from a point downstream of the reactor back to the reactor feed stream provides a reduction in raw materials consumption per tonne of product. This has both economic and environmental benefits. There are also benefits in avoiding the treatment or disposal of unreacted feed as a waste or by-product from the process.

There are, however, economic and environmental burdens in recycling unreacted feed due to

- separation of unreacted feed from the product and possibly from other molecular species in the reactor outlet stream
- pumping, compression, or conveying of unreacted feed back to the reactor inlet; this is particularly important where a high pressure reaction is followed by a low pressure separation.

Such economic and environmental burdens must be weighed up against the benefits of recycling. Processes where unreacted feed is recycled include chlorine, ethylene, and PVC manufacture.

Reactant conversion can also be limited in some batch reactors due to a decrease in reaction kinetics as conversion proceeds. A point in the batch reaction cycle is reached where conversion rate is judged too slow to be economically viable. There is then the task of separating the desired product from the reactant and recycling the unused reactant for the next batch. An example occurs in the batch polymerisation of vinyl chloride monomer to PVC.

## 4.8 Reversible Reactions

Where reactions are reversible and one reactant is a pollutant, difficulties are created for waste minimisation since reactant conversion is limited by the reaction equilibrium. An example is the reversible oxidation of $SO_2$ to $SO_3$ in sulphuric acid manufacture.

$$SO_2 + 1/2O_2 \rightleftharpoons SO_3 \qquad \Delta H = -298.3 \, \text{kJ/mol}$$

Reactant conversion can often be improved by selecting temperature, pressure and concentration conditions so that equilibrium favours the forward reaction.

There are *general rules for assisting forward direction of reversible reactions*. These rules follow Le Chatelier's principle.

**Reactor temperature**    should be increased for an endothermic reaction, and decreased for an exothermic reaction.

**Inert concentration**    If reaction involves a decrease in number of moles, inert concentration should be decreased. If reaction involves an increase in number of moles, inert concentration should be increased.

**Reactor pressure**    Vapour phase reactions with a decrease in number of moles should be set at higher pressures. Vapour phase reactions with an increase in number of moles ideally should have a pressure continuously decreasing as conversion proceeds. Practical considerations such as cost and use of pressure vessels, piping, and compressors, as well as safety and operability issues, sometimes restrict choices of reactor pressure.

Reactant conversion in the overall process can be increased by providing additional stages of reaction. This, however, incurs increased energy consumption derived from increased pressure drop over the reaction system, as well as increased capital costs associated with the larger reactor and any ancillary equipment and piping.

An important aspect of reversible exothermic reactions is the dependence of conversion and kinetics on temperature. Often a balance is required between choosing a high temperature where kinetics is favourable and a lower temperature where thermodynamic equilibrium is favourable. This is the case with the oxidation of $SO_2$ to $SO_3$ where a lower temperature would favour the equilibrium but the reaction kinetics are too slow below around 430°C.

Another good example of this need for balance between kinetically and thermodynamically favourable conditions is the shift reaction in the production of hydrogen, where unreacted carbon monoxide produced in the upstream steam reformer is reacted with steam in the presence of a catalyst.

$$H_2O + CO \rightleftarrows H_2 + CO_2 \qquad -41 \text{ kJ/mol}$$

High temperature and low temperature shift reactors are used commercially with separate catalysts and separate heat recovery systems (Twigg, 1989).

## 4.9 Catalysis

Catalysts have an important role in improving conversion and selectivity, and also where high reactor temperatures and pressures are involved, in allowing reactions to occur at lower temperature and pressure. Lower temperatures and pressures reduce energy consumption in heating and in pumping (or compression) duties. Lower temperatures can also reduce the severity of attack by process fluids on materials of construction in the reactor and associated plant. Lower temperatures and pressures can also reduce safety risks when handling hazardous chemicals.

Catalysts typically degrade during their life and must be replaced when catalyst performance is judged to be unsatisfactory. Catalyst replacement usually requires a plant shut-down for removal of spent catalyst and installation of new catalyst. Catalyst degradation has a number of effects on process performance:

- deterioration in reactor performance (in one or more of conversion, yield, selectivity)
- possible increased pressure drop through fines generation in the case of solid catalysts
- the need for catalyst regeneration or disposal; disposal is a problem if the catalyst contains toxic species such as heavy metals.

### 4.9.1 *Example of the Effect of Catalyst Activity on Performance*

An example of the effect of catalyst activity on reactor performance is the quality of by-product hydrogen from a catalytic reformer unit (CRU) in a petroleum refinery. The purpose of the CRU is to increase octane number of gasoline, mainly by dehydrogenation, isomerisation, and cyclisation of hydrocarbon feed. Hydrogen is a by-product of each hydrocarbon reaction; for example, in cyclisation of *n*-hexane where

$$C_6H_{14} \rightarrow C_6H_6 + 4H_2$$

The catalyst used is in the CRU is platinum supported on a silica or silica alumina base. Note the lower hydrogen content and increased hydrocarbon content of gas as the catalyst ages.

| | Gaseous composition (% v/v) | |
| --- | --- | --- |
| Component | Middle of run | End of run |
| $H_2$ | 86.0 | 77.0 |
| $CH_4$ | 6.5 | 13.5 |
| $C_2$–$C_5$ | 7.5 | 9.5 |

## 4.10  Agent Materials

Agent materials have advantages in enabling the desired reaction to proceed efficiently, but disadvantages if they have toxic properties or indirectly cause waste formation. Examples of agent material functions include

- solvents
- reaction enablers
- reaction stoppers.

Agent materials may contaminate process streams, products and effluents. Examples of applications of agent materials in reactors include

- use of mercury in chlor-alkali production. This causes a range of environmental problems due to toxicity of mercury through product and effluent contamination.
- use of a reaction stopper to stop a runaway reaction in PVC reactor. While successful in halting the runaway, the batch of PVC becomes contaminated and unfit for sale.

## 4.11  Case Examples

Table 4.2 shows some common chemical processes where sources of waste generation in reactors and waste minimisation strategies previous discussed are seen to apply.  The processes are then discussed in this context. The

Table 4.2   Waste sources and waste minimisation strategies in selected chemical process examples

| Waste source | Strategy | Hg cell $Cl_2/NaOH$ | $C_2H_4$ | HCN, NaCN | $H_2SO_4$ | PVC |
|---|---|---|---|---|---|---|
| Feed impurity | | **** | | **** | **** | |
| | Purify feed | **** | | **** | **** | |
| Incomplete conversion | | **** | **** | **** | **** | **** |
| | Recycle feedstock | **** | **** | **** | | **** |
| Secondary reactions | | **** | **** | **** | | |
| Reversible reaction | | | | | **** | |
| Catalyst degradation | | | | **** | **** | |
| Agent materials | | **** | | | | **** |
| | Static mixer | | **** | **** | **** | |
| | Waste heat boiler | | **** | **** | **** | |
| | Direct quench | | **** | | | |

**Note:** **** denotes that the waste source or strategy is applicable for the particular process.

processes also provide useful examples of waste minimisation in separation processes, energy use, materials recycling, and plant operations referred to in later chapters of Part B.

## 4.12 Chlor-Alkali Production in Mercury Cell

Figure 4.1 shows the reaction system for a chlor-alkali plant using mercury cells to produce chlorine, caustic soda, and hydrogen. The reaction system comprises multiple mercury cells arranged in parallel to achieve the required plant production capacity. Each mercury cell comprises two reactors, a brine cell, and a soda cell. Direct current electricity is supplied through a transformer rectifier to the brine cell. The chemical reactions occurring in each cell are as follows:

**Brine cell reactions:** $Na^+ + e^- \xrightarrow{Hg} Na/Hg$

$$Cl^- \longrightarrow 0.5Cl_2 + e^-$$

**Soda cell reaction:** $Na/Hg + H_2O \longrightarrow NaOH + 0.5H_2$

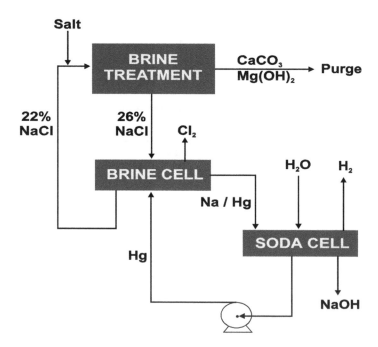

**Figure 4.1** Simplified flow diagram for a mercury cell reaction system.

Mercury acts as a cathode in the brine cell and flows along the steel base of the rectangular cell in which titanium anodes are set. The sodium amalgam leaves the brine cell and flows to the soda cell where water is added to form caustic soda and hydrogen. The sodium-free mercury is pumped back to the brine cell. Sodium chloride conversion is incomplete in the brine cell; depleted brine is dechlorinated by vacuum or air stripping, and reconcentrated by fresh salt addition. Calcium and magnesium impurities in the brine are removed by precipitation, gravity settling and filtration. Calcium is removed as calcium carbonate and magnesium as magnesium hydroxide; the resulting insoluble salts are removed with some brine as a purge stream.

## 4.12.1 *Transport Paths*

Mercury exists in three forms: elemental mercury, inorganic mercury compounds (primarily mercuric chloride), and organic mercury compounds (primarily methyl mercury). Each form is toxic, and exhibits distinct health effects. Mercury is a pollutant of concern due to its persistence in the environment, potential to bio-accumulate, and toxicity to humans and the environment. Useful references on the environmental properties of mercury can be found at http://www.epa.gov/ttn/atw/hlthef/mercury.html and www.osha.gov/SLTC/mercury/index.html.

It is important to identify the potential paths of a toxic species from a process to the environment. In the case of mercury in chlor-alkali plants, contamination of products such as caustic soda or hydrogen implies potential contamination for downstream processes and products. Contamination of other process streams and hence effluents, as well as emissions to the environment due to loss of containment, are also of concern. Transfer paths for mercury to the environment include

(i) from brine cell:
- leakage from equipment and piping handling the recycle brine stream
- purge containing $CaCO_3$ and $Mg(OH)_2$ from the brine circuit
- gas stream leaving the dechlorinator in the return brine circuit prior to alkaline scrubbing (of waste chlorine streams)
- spillage and evaporation during periodic cell maintenance and cleaning

(ii) from soda cell:
- leakage from mercury pump seal failure and piping in mercury reticulation

- contamination of caustic soda product
- contamination of hydrogen product
- contamination of plant by-products, for example, hydrochloric acid (from hydrogen and chlorine) and sodium hypochlorite (from chlorine and caustic soda)
- spillage and evaporation during periodic cell maintenance and cleaning.

Mercury can be largely removed from the caustic soda by filtration through activated carbon, and from the hydrogen stream by adsorption onto carbon impregnated with sulphur. In both these cases, however, there remains the problem of either

- disposal of the carbon residues contaminated with mercury or
- attempting to recover the mercury from the carbon residues.

The *preventive solution* lies in replacing mercury cell technology with membrane cell technology. The membrane, shown in Fig. 4.2, allows passage of sodium ions but not chloride ions or hydroxide ions. By eliminating the environmentally harmful mercury species from the process, all mercury waste streams are eliminated. Membranes are separation devices increasingly used in the process industries in a wide range of applications, including those achieving environmental improvement. Membranes are an example of innovative technology undergoing rapid change in recent years. The nature, performance characteristics, and configurations of membranes in separations, as well as related performance modelling, are discussed by Smith (2005).

Mercury cell chlorine plants are gradually being replaced by membrane plants worldwide. In 2007, mercury-based plants accounted for 37.7%

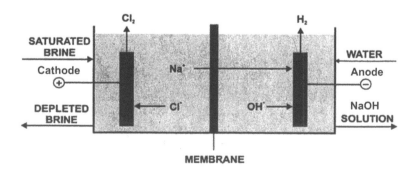

**Figure 4.2**  Schematic diagram of membrane cell.

of European chlorine capacity; 42 mercury-based plants remained to be converted or phased out by 2020 at a cost of EUR 30,000 million. (http://www.eurochlor.org).

Mercury is present in many fossil fuels including coal, natural gas and petroleum, and minerals such as bauxite and zinc sulphide ores. While the concentrations of mercury in these cases may be small, the effective removal of mercury and avoiding its transfer to the environment is a challenging task. Mercury sources and emissions are discussed in the context of industrial ecology by Allen (2004). A case problem on utilisation of by-product hydrogen from chlorine plants involving mercury removal is provided in Part D of this book.

## 4.12.2 *Other Aspects of the Mercury Cell Chlorine Process*

Apart from use of a toxic agent material, there are other examples of reactor waste sources in chlor-alkali production.

*Secondary reactions* can occur in the brine cell, involving hydrogen production. At >3% hydrogen in chlorine gas, a gas explosion can occur. Hydrogen formation in the brine cell can be reduced by *brine purification*, especially by limiting calcium and magnesium concentrations. Raw salt quality can be improved by washing immediately after harvesting. Typical salt composition from Dampier, Western Australia contains approximately 99.75% NaCl, 0.04% calcium, and 0.025% magnesium by weight on a dry basis. The removal of calcium and magnesium from brine involves some salt losses and (in the case of mercury cells some mercury losses) through sludge or filter cake removal.

There is also *limited conversion* (around 15%) of feed brine ($\sim$26% NaCl) to chlorine in the chlorine cell; the depleted brine ($\sim$22% NaCl) is *recycled*, resaturated, and purified.

## 4.13 Ethylene Manufacture from Hydrocarbons

Ethylene can be manufactured from paraffin hydrocarbon feedstocks, ranging from ethane, through LPG and naphtha, to gas oil. As the molecular weight of the feedstock increases, the tonnages of co-products (propylene, butadiene, petrol) increases. The capital cost of the plant also increases, principally due to increased separation and handling of the larger co-product quantities.

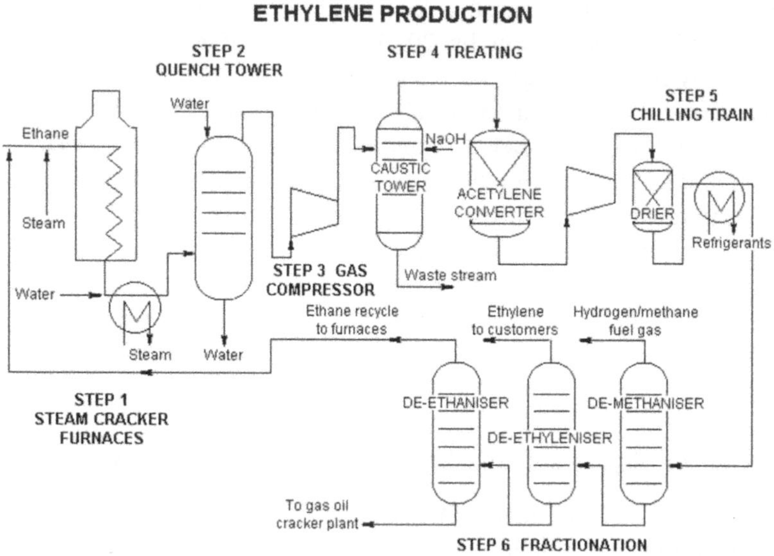

**ETHYLENE PRODUCTION**

**Figure 4.3** Simplified flow diagram for making ethylene from ethane. Reprinted with permission from Qenos.

Figure 4.3 shows a simplified flow sheet for ethylene production from ethane. The reaction in the furnace reactor has a short residence time (<1 second). Steam is added as a diluent, reducing the hydrocarbon concentration, to increase ethylene yield and minimise carbon deposition in the reactor. A conversion of ethane between 60% and 70% by weight is typically achieved. Rapid cooling of the reaction gases is necessary to minimise side reactions.

A typical once-through yield pattern for a steam cracking furnace with ethane, butane, or naphtha feed is shown in Table 4.3. The yield pattern indicated in the table is based on a range of data extracted from Ullmann (2002), Kirk and Othmer (1994), Mouljin, Makkee, and Van Diepen (2001), and the Qenos website. Once-through yield of ethylene using ethane feed approximates 52% by weight. By recycling unreacted ethane, assuming all ethane is cracked to extinction, ethylene yield typically approximates 77% by weight.

The key process steps can be summarised as

- high temperature reaction (approximately 900°C) where ethane in the presence of steam is converted in a furnace reactor to ethylene and hydrogen $C_2H_6 \rightarrow C_2H_4 + H_2$
- rapid cooling of reaction products with waste heat recovery enabling high pressure steam generation

Table 4.3 Typical once through yield pattern for ethane, butane, and naphtha feedstocks (mass basis)

| | Feedstock | | |
|---|---|---|---|
| **Product** | **Ethane** | **Butane** | **Naphtha** |
| Hydrogen | 4 | 1.1 | 0.94 |
| Methane | 3.5 | 19.9 | 14.7 |
| Acetylene | 0.5 | 0.5 | 0.52 |
| Ethylene | 52.1 | 37.0 | 29.1 |
| Ethane | 35.0 | 3.7 | 3.7 |
| Propylene | 1.1 | 17.1 | 15.9 |
| Butadiene | 1.7 | 4.2 | 4.9 |
| Butene | 0.2 | 4.3 | 4.6 |
| Butane | 0.2 | 6.1 | 0.5 |
| Benzene | 0.7 | 2.5 | 6.8 |
| Other $C_x H_y$ | 0.5 | 1.0 | 7.1 |
| Gasoline | 0.3 | 1.8 | 8.1 |
| Fuel oil | 0.2 | 0.8 | 3.1 |
| **Total** | 100 | 100 | 100.0 |

- quenching the cooled gases leaving the waste heat boiler by direct contact with water
- compression and purification to remove $CO_2$, $H_2S$, $C_2H_2$, and water
- chilling by refrigeration, usually using ethylene and propylene refrigerants available from the plant
- a distillation sequence to separate methane/hydrogen, ethane, ethylene, and heavier hydrocarbons
- *recycle of separated ethane* to the reactor feed.

## 4.14 Hydrogen Cyanide Manufacture from Ammonia, Methane, and Air

Hydrogen cyanide (HCN) is manufactured by the reaction of methane with ammonia and oxygen in the Andrussow process. HCN is an intermediate for making a number of chemicals. One key chemical is sodium cyanide, used for the extraction of gold from ores.

The main reaction occurs around 1150°C over a platinum rhodium catalyst

$$CH_4 + 1.5O_2 + NH_3 \rightarrow HCN + 3H_2O - 481.9 \text{ kJ/mol}$$

Table 4.4   Estimate of post reactor gas concentrations for Andrussow process

| Component | Concentration by volume |
|-----------|-------------------------|
| HCN | 7.8% |
| $NH_3$ | 2.3% |
| $H_2$ | 15.0% |
| $N_2$ | 49.0% |
| $CH_4$ | 0.5% |
| CO | 3.9% |
| $H_2O$ | 21.1% |
| $CO_2$ | 0.4% |

A static mixer is used upstream of the reactor to ensure uniform gas concentration and temperature of the reactor feed and also to avoid any regions of flammable mixtures in the reactor feed.

Raw materials conversion is incomplete and some secondary reactions occur. Estimates in Ullmann (2002) and Kirk and Othmer (1994) indicate small concentrations of unreacted ammonia and methane, significant concentrations of hydrogen, and smaller concentrations of carbon monoxide and carbon dioxide present in gas leaving the reactor. Hydrogen is present because of incomplete oxidation of the hydrogen present in the methane and ammonia feedstocks. Table 4.4 provides an indicative composition of post reactor gas based on estimates drawn from Ullmann and Kirk and Othmer.

Reaction is followed immediately by rapid cooling of gases in a waste heat boiler. The rapid cooling is undertaken to prevent hydrogen cyanide decomposition, but allows process heat recovery with steam generation. Some literature flow sheets show recycling of unreacted ammonia through an absorption stripping process, as indicated on Fig. 4.4. Finally, HCN is absorbed in

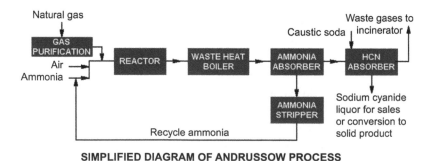

SIMPLIFIED DIAGRAM OF ANDRUSSOW PROCESS

**Figure 4.4**   Simplified diagram of Andrussow process for HCN manufacture.

caustic soda solution, where sodium cyanide is the product, or alternatively in water for subsequent stripping, where other products are manufactured.

**Feed purification**    Natural gas is the usual methane source for the Andrussow process. Natural gas is normally distributed by pipeline as sales gas for generalised industrial, commercial, and domestic use. A typical composition by volume of sales gas is

90.6% $CH_4$
5.6% $C_2H_6$
0.8% $C_3H_8$
0.2% $C_4H_{10}$
1.1% $N_2$
1.7% $CO_2$.

Even when meeting sales gas specification for natural gas, the natural gas feed needs to be purified to remove heavier hydrocarbons $(C_2{}^+)$, carbon dioxide, and small quantities of sulphur prior to the main reaction. Sulphur is removed to avoid poisoning the catalyst which would lower conversion rates. Heavier hydrocarbons are removed to reduce carbon deposition on the catalyst. $CO_2$ is removed to avoid formation of sodium carbonate during absorption of the HCN in alkaline solution. Sodium carbonate is difficult to separate from sodium cyanide and becomes a product impurity as well as consuming additional sodium hydroxide. Air, as the source of oxygen, is filtered to protect the reactor catalyst from air-borne particles.

## 4.15  Sulphuric Acid Manufacture

In some cases where reactant conversion is limited, reactant recycle is not viable. An example occurs in sulphuric acid manufacture, where separation of $SO_2$ (from $N_2$) and its subsequent recycle would be both complex and costly. Sulphuric acid manufacture is an example of a process with a *reversible reaction* where one reactant, sulphur dioxide, is a pollutant. In this case, reliance is placed on increased reactant conversion without resorting to recycle.

Main reactions involve the reversible oxidation of sulphur dioxide to sulphur trioxide:

$$SO_2 + 0.5O_2 \rightleftharpoons SO_3 \qquad \Delta H = -298.3 \text{ kJ/mol}$$

and the subsequent reaction of sulphur trioxide with water which occurs immediately following the absorption of sulphur trioxide in 98% sulphuric acid:

$$SO_3 + H_2O \rightarrow H_2SO_4 \qquad \Delta H = -130.4 \text{ kJ/mol}$$

Commercial feedstocks for sulphuric acid manufacture are sulphur, or alternatively sulphur dioxide formed during smelting of mineral sulphide ores. Figure 4.5 shows a simplified process flow diagram for manufacture of sulphuric acid from sulphur. Where sulphur is the feedstock, heat is liberated from the combustion of molten sulphur in air within a sulphur burner. The temperature of the sulphur dioxide containing gases ($SO_2$, $N_2$, $O_2$) leaving the sulphur burner is around 1000°C; these gases are cooled in a waste heat boiler to around 430°C for feeding to the first reaction stage. The reactor in sulphuric acid plants is often referred to as a 'converter'.

Reactor temperature control and cooling between reaction stages allows

- optimal inlet temperatures to reaction stages (~430°C)
- waste heat recovery with steam generation (sulphur feed) or feed preheat (metallurgical source of $SO_2$).

Reactor temperature and reactant concentration control have sometimes been achieved using air dilution, providing both cooling and increased oxygen concentration.

While higher pressure would favour the equilibrium of the forward $SO_2$ oxidation reaction, low pressure plants are generally adopted for sulphuric acid for the practical advantage of avoiding pressure vessels.

Product removal between reaction stages by intermediate absorption of $SO_3$ favours equilibrium for the forward $SO_2$ oxidation reaction at the final reaction stage. The result is a greater overall conversion of $SO_2$ to $H_2SO_4$. Absorption occurs most favourably at an acid inlet temperature of 80°C and gas inlet temperature of 220°C.

Increasing the number of reaction stages increases the overall conversion of $SO_2$ to $SO_3$; reaction stages have been increased in more recent plants from 4 to 5.

High pressure steam generated in a sulphur burning plant can be used for generating electricity which exceeds the plant's requirements; surplus electricity can be exported to the grid. Low pressure steam can be used for heating within the plant, for example, in melting sulphur for feeding to the sulphur burner.

Other features of sulphuric acid plants include

- integrated acid circulation circuits with concentrated sulphuric acid used in both drying and absorption columns; moisture in incoming air contributes to water make-up for raw material needs

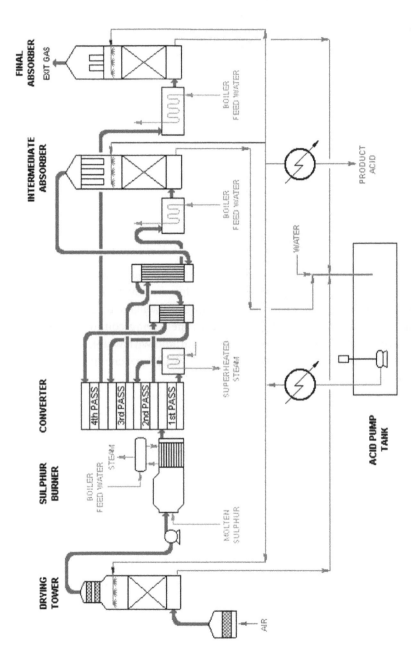

**Figure 4.5** Simplified flow diagram for sulphur burning sulphuric acid plant. Reprinted by permission of Simon-Carves.

- provision of entrainment and mist removal devices
- layer of packing to remove entrained liquid in drying column
- demisters to remove mist in absorption column
- heat removal in drying and absorbing acid circuits, reflecting the exothermic nature of acid dilution (drying) and $SO_3$ absorption in water (absorbing).

**Metallurgical plants**   The flow sheet shown in Fig. 4.5 is for a sulphur burning plant. An alternative and common source of $SO_2$ is the smelting or roasting of sulphide ores (e.g., Pb, Zn, Cu, Ni). Downstream of the ore roaster and upstream of the acid plant,

- heat is recovered from the combustion of the metal sulphide and used for steam generation
- dust is recovered (for its mineral value)
- gas is further cooled and cleaned prior to feeding to the acid plant.

Gas cleaning comprises aqueous scrubbing followed by wet electrostatic precipitators. Gas cleaning is important for removal of heavy metal compounds derived from the mineral sulphide ore which would contaminate the product acid. After cleaning, gas enters the drying column of the acid plant at around ambient temperature and saturated with water. After drying, the gas is heated to reactor temperature ($\sim 430^\circ$C) by exchanging heat with hot gas leaving the reactor.

Aspects of sulphuric acid manufacture are further discussed in Chapters 9 (in the context of plant start up) and 14 (in the context of enviroeconomic assessment). More detailed discussion of sulphuric acid manufacturing technology is provided in Mouljin *et al.* (2001), Twigg (1989), and Ullmann (2002).

## 4.16  PVC Production by Suspension Polymerisation of Vinyl Chloride Monomer

Polymerisation of vinyl chloride monomer (VCM) to produce poly-vinyl chloride (PVC) is a batch process. Figure 4.6 shows a simplified flow sheet for PVC manufacture. VCM and water are charged to the reactor. A protective colloid is added as a suspending agent. The vessel contents are agitated, dispersing the VCM as liquid droplets within the water. The mixture is then heated to the required temperature for reaction, and the polymerisation initiated using a free radical initiator. The reaction is exothermic and

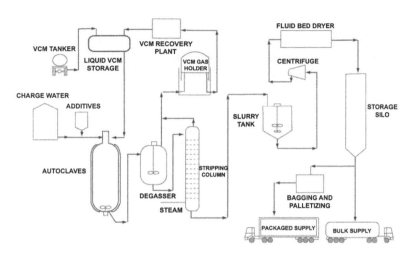

**Figure 4.6** Simplified process flow diagram of PVC plant. Reprinted with permission from Australian Vinyls Corporation.

heat is removed from the reactor, commonly by cooling water circulating through an external jacket; in some cases additional heat transfer surfaces may be provided. As the batch reaction proceeds and conversion of VCM progressively increases, the reaction rate slows. It is generally uneconomic to proceed beyond a conversion of around 90%. At this point the batch reaction is terminated.

Unconverted VCM is then separated and recovered for use in subsequent batches. Recovery is achieved by first degassing the polymer solution which is discharged from the reactor under reduced pressure, followed by stripping residual VCM from the polymer in a steam stripping column. VCM is then separated from water, and recovered by compression and refrigeration. PVC from the steam stripping column is passed through a centrifuge and then a dryer to achieve the required moisture content for the PVC product. The PVC product, either in bulk or packaged form, is finally transported for distribution to users.

The process is of interest in several aspects. Firstly, it is an example of a *batch reaction process*, where the reaction batches are sequenced to enable continuous processing for VCM recovery and PVC dewatering and drying.

Secondly, it is a case where *conversion is limited* by reaction kinetics and economic considerations; this requires additional processing of VCM, a known carcinogen and flammable material. Any unrecovered VCM in the vent stream from the liquefaction plant is normally incinerated.

Thirdly, the exothermic nature of the reaction provides *risk of a runaway reaction.* Should a runaway reaction occur, a reaction stopping agent, *4-tertiary-butyl-catechol,* removes the free radicals necessary for polymerisation. Addition of this *agent,* however, renders the PVC from the batch unsaleable.

Fourthly, the process provides an example of the need to clean a reactor between batches. In the case of the PVC reactor, the cleaning should be carried out on a sealed reactor to avoid exposure of plant personnel to VCM. There is need to rinse out residual PVC from the reactor with flush water, since residual PVC would

- lead to a downgraded product in a subsequent batch
- prevent effective coating of reactor internals with build-up suppressant (necessary to avoid accumulated deposit of PVC solid during the batch).

Finally, the process is a *large consumer of water*, mainly through additions in the reactor and through evaporative losses from cooling water towers. Water consumed in an operating plant could be as high as 5 tonne per tonne of PVC.

## 4.17 Concluding Remarks

The five cases discussed show examples of waste generated in reactors within chemical processes, and steps to minimise such waste. Some implications of the generation of waste in the reactor for the complete process have also been identified. These cases are broadly indicative of reactor waste generation and its implications in the wider spectrum of industrial chemical processes.

## References

Allen, D. T. (2004) An industrial ecology: material flows and engineering design, in *Sustainable Development in Practice* (ed. Azapagic, A., Perdan, S., and Clift, R.), Wiley, Chichester, UK.

Australian Vinyls Corporation. http://www.av.com.au/.

Kirk, and Othmer. (1994) *Encyclopaedia of Chemical Technology*, 4th edn, Wiley, New York.

Mouljin, J. A., Makkee, M., Van Diepen, A. (2001) *Chemical Process Technology*, Chapter 4, Wiley, Chichester, UK.

Qenos Australia. http://www.qenos.com/.

Sheldon, R. A. (1994) Consider the environmental quotient, *Chemtech*, Vol. **24(3)**, 38–47.

Simon-Carves. *Sulphuric Acid Plants*, Company brochure.

Smith, R. (2005) *Chemical Process Design and Integration*, Wiley, Chichester, UK.

Twigg, M. (ed.) (1989) *Catalyst Handbook*, 2nd edn, Wolfe Publishing, London, UK.

Ullmann. (2002) *Encyclopaedia of Industrial Chemistry*, Wiley-VCH, Weinheim, Germany.

# Chapter 5

# Waste Minimisation in Separation Processes

## 5.1 Classification of Separation Processes

The separation of different materials is an integral part of all industrial processes. Separation processes are required to purify naturally occurring minerals and fossil fuels prior to their use as feedstocks and fuels. Separation processes are required downstream of chemical reactors, to ensure desired products are separated from unreacted feedstocks and undesired side products. Separation processes are widely used to achieve required product quality, and to ensure any emissions to the environment meet regulatory standards. Incomplete separation can have adverse consequences within a process, in a product manufacturing chain, and in product use.

Separation processes can involve

- separation of distinct phases, for example, liquid from solids by settling, filtration, drying
- separation of components from a gas stream, for example, by gas absorption, pressure swing adsorption, membrane separation
- separation of components from a liquid stream, for example, by evaporation, distillation, membrane separation.

Separation processes are sometimes classified by their application to *heterogeneous* (multiphase) or *homogeneous* (single phase) mixtures (Smith, 2005). For homogeneous mixtures, separation can be achieved by

*Sustainable Process Engineering: Concepts, Strategies, Evaluation, and Implementation*
David Brennan
Copyright © 2013 Pan Stanford Publishing Pte. Ltd.
ISBN 978-981-4316-78-1 (Hardcover), 978-981-4364-22-5 (eBook)
www.panstanford.com

- creating a separate phase through
  - ▶ vaporisation or boiling
  - ▶ condensation
  - ▶ crystallisation
- vapour–liquid separation through
  - ▶ distillation
  - ▶ evaporation
- using extraneous materials as mass separating agents in
  - ▶ absorption
  - ▶ stripping
  - ▶ liquid–liquid extraction
  - ▶ adsorption
  - ▶ drying.

## 5.2 Sources of Waste in Separation Processes

Wastes are frequently generated in separation processes

- directly from imperfect separations
- indirectly through consumption of energy or utilities
- from use of extraneous materials such as solvents, adsorbents, filter aids, membranes.

Wastes are generated in normal operation in accordance with design intent, but also in abnormal operation, and in start-up and shut-down procedures.

The sources of waste generation in separation processes must be identified before taking steps to minimise the waste. Different separation methods typically generate waste in distinct ways, as discussed below. Apart from these waste sources characteristic of specific separation methods, energy consumption is common to all separation equipment. Energy may be consumed *directly* as a utility, or *indirectly* through pressure loss due to flow through equipment which contributes to compression or pumping energy. Energy or utility consumption implies a spectrum of impacts through resources consumed and wastes emitted in providing the utility. Resource consumption and waste generation in utility systems are explored in Chapter 6.

Some main sources of waste are now identified in selected unit operations. The term 'utility wastes' denotes that a range of wastes are generated attributable to the use of a specific utility.

## 5.3 Distillation

Distillation relies on the use of heat to separate two or more components of different relative volatility. Distillation is carried out in a vapour–liquid contacting column which may be packed or tray. Figure 5.1 shows a simplified diagram for the separation of two components by distillation in a tray column.

Potential sources of waste generation in distillation include

- characteristic utility wastes derived from utility use in the condenser and reboiler, and in the pumping of feed and reflux streams
- accumulation of non-volatiles in the reboiler
- accumulation of non-condensables in the condenser
- fouling of heat exchangers and distillation column internals
- out of specification product through loss of control or abnormal operation
- possible contamination of a heating or cooling utility, or a process stream through heat exchanger leakage
- aqueous effluents contaminated with organics resulting from live steam injection (live steam injection is an alternative to using a reboiler where aqueous mixtures are distilled)
- contamination of streams resulting from use of entrainers in azeotropic distillation.

Energy is consumed through the use of utilities in the following aspects:

| Energy use | Typical utility | Wastes |
|---|---|---|
| Reboiler | Steam, fuel | Utility wastes |
| Condenser | Cooling water | Utility wastes |
| Pumping (feed, product, reflux) | Electricity | Utility wastes |

Energy consumption is influenced by the degree of difficulty in separation. For a single distillation column, a number of design decisions relate to the ease of separation. One example is the choice of reflux ratio, the ratio of flow rate of overhead condensate returned to the top of the column to the flow rate of overhead product. In selecting the reflux ratio, there is a trade-off between capital cost and energy consumption. As the reflux ratio is increased, the number of trays required is reduced, reducing the height of the column but marginally increasing its diameter, and increasing the utility requirements for both boiling and condensing duties.

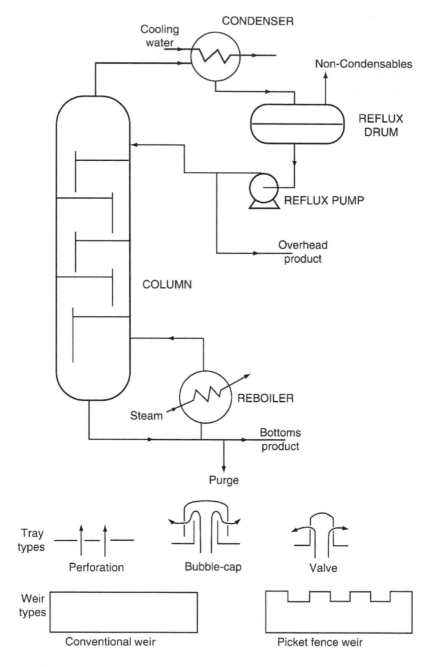

**Figure 5.1** Binary distillation tray column with ancillary equipment and alternative tray and weir designs. Weirs enable retention of liquid on trays. Trays enable vapour-liquid contact for mass transfer.

Trade-offs also exist between the extent of separation and cost. In these cases, the process designer should also consider the costs of any effluent treatment or disposal and the associated environmental burdens.

## 5.4  Gas Absorption

Gas absorption relies on the relative solubility of one component of a gas stream (the solute) in a liquid absorbent (the solvent). A simplified diagram for a gas absorption system is shown in Fig. 5.2.

Gas absorption, like distillation, occurs in a vapour–liquid contacting column. The absorption of the gas in the liquid solvent may be physical (e.g., absorption of acetone vapour in water) or may be accompanied by a chemical

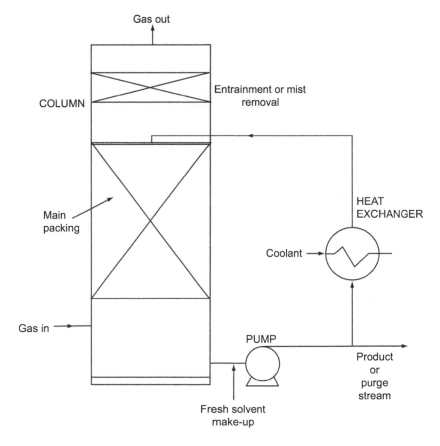

**Figure 5.2**   Gas absorption in packed column with external cooling.

reaction (e.g., absorption of chlorine gas in aqueous sodium hydroxide to form sodium hypochlorite). In the system shown in Fig. 5.2, gas absorption occurs in a packed column and is exothermic, with heat removed externally in the solvent recirculation loop. Fresh solvent is added to the circulation loop, and a product or purge stream removed.

Possible sources of waste generated in gas absorption operations include

- effluent gas contamination by unabsorbed species (incomplete absorption)
- solvent entrainment in gas leaving the absorption column
- purge from circulating solvent where the purge has no market
- build up of contaminants in recirculating solvent; contaminants could be introduced from the gas stream, or could accumulate as corrosion products.

The extent of absorption can be increased by lowering the temperature and increasing the pressure in the absorption column. These initiatives, however, typically involve increased energy consumption which must be considered. Energy consumption is also influenced by column pressure drop, which increases with the increased packing height and column height required for more difficult absorption duties. Column internals other than packing also contribute to pressure drop in the column.

Energy is consumed in gas absorption arising from gas and liquid transport, apart from any heat removal considerations derived from exothermic absorptions.

| Energy use | Typical utility | Wastes |
|---|---|---|
| Gas transport | Electricity | Utility wastes |
| Liquid pumping | Electricity | Utility wastes |

Additional energy is consumed in solvent regeneration for those systems involving absorption and subsequent stripping of solvent to recover the absorbed gas. In such cases, the stripping column acts as a distillation column, and involves energy consumption and wastes inherent in distillation. Figure 5.3 shows a combined absorption stripping system for solvent recovery from a gas stream. In this example, live steam is used, avoiding the cost of a reboiler; a downside of using live steam in this case is the creation of an aqueous effluent resulting from the resulting water imbalance.

Some case examples of gas absorption are discussed later in this chapter.

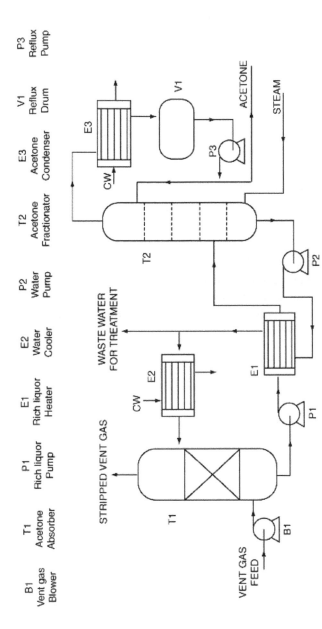

**Figure 5.3**  Simplified flow diagram for absorption-stripping plant used for acetone recovery.

## 5.5 Adsorption

Adsorption involves separation of components by adsorbing one component, *the adsorbate* onto the surface of a solid material, *the adsorbent* (Smith, 2005).

Adsorption is commonly used for separating components in gases, but can also be applied for separating components in liquids. Adsorption can occur as

- physical adsorption, where physical bonds are formed between adsorbent and adsorbate
- chemical adsorption, where chemical bonds are formed between adsorbent and the adsorbate.

Common examples of adsorbents include

- activated carbon, a highly porous solid made from a range of carbonaceous materials
- silica gel, a porous amorphous form of silica
- activated alumina, a porous solid with high surface area
- molecular sieve zeolites, crystalline aluminosilicates where adsorption occurs within the crystals.

Applications of physical adsorption include

- separating hydrogen from hydrocarbons in the hydrogen rich by-product of a catalytic reforming unit
- separating hydrogen from carbon dioxide in hydrogen manufacture by steam reforming of methane
- separating water vapour from hydrocarbons in ethylene manufacture
- recovering organic solvents from gases.

Adsorption is normally carried out using several beds operating in a sequence. When the adsorptive capacity of adsorbent in one bed is spent, the adsorbate is released by pressure reduction (pressure swing) or temperature increase (temperature swing) or use of steam. Effective recovery of adsorbate from loaded adsorbent is an important consideration in the regeneration stage. This often involves condensing the regeneration stream to recover the adsorbate. Where the adsorbate is an organic solvent, separation of condensed solvent from water is required often involving a vent stream and some contamination of the water.

Applications of chemical adsorption include removal of mercury vapour from natural gas, or from hydrogen produced in mercury cell chlorine plants, where carbon impregnated with sulphur is often used as the adsorbent. Adsorbed mercury reacts with sulphur to form mercuric sulphide.

Possible sources of waste in adsorption include

- incomplete adsorption of the adsorbate during the adsorption cycle
- incomplete recovery of adsorbate from the regeneration stream during regeneration
- disposal of spent adsorbents.

Energy can be consumed in adsorption in a number of aspects.

| Energy use | Utility | Waste |
|---|---|---|
| Fluid transport through bed | Electricity | Utility wastes |
| Regeneration | Electricity, fuel or steam | Utility wastes |
| Chemical recovery from regeneration stream | Cooling water | Utility wastes |

## 5.6 Filtration

Filtration is commonly used to separate a solid from a liquid. A wide range of filtration equipment is available. Filter aid materials are sometimes added, either as precoat to the filter medium, or as body feed, to aid the flow of liquid through the retained solid (or 'filter cake'). Separated solid and liquid streams may both have economic values, but often either the solid or liquid stream has no market and is a waste. Characteristic sources of waste in filtration include

- cake washing, which can lead to contamination of wash liquor by impurities from the filter cake
- disposal of filter aid
- treatment and disposal of solid material from the filtration where no market exists for solid product
- filtrate contamination through filter medium failure
- liquid material separated from the filtration, where no market exists for the liquid.

Energy is consumed in moving the solid–liquid suspension through the filter medium.

| Energy use | Utilities | Wastes |
|---|---|---|
| Pumping | Electricity | Utility wastes |
| Vacuum (in some cases) | Electricity or steam | Utility wastes |

### 5.6.1 *Centrifugal Separation*

Centrifuges are an alternative to filters for solids–liquid separations. Wastes are characteristically generated by

- solids washing
- spillage.

Energy is consumed in conveying liquid and solid materials and in the high speed rotation of the centrifuge.

| Energy use | Utilities | Wastes |
|---|---|---|
| Centrifugal action | Electricity | Utility wastes |
| Liquid, solids conveying | Electricity | Utility wastes |

If not suppressed, *noise* generated by high speed (up to 10,000 rpm) can prove an environmental or occupational safety impact.

### 5.6.2 *Filtration of Solids from Gas Streams*

Particulate solids are often present in gas streams, as a result of entrainment from other operations such as solids drying. Cyclones, which involve no moving parts, effect separation by tangential gas entry creating a centrifugal field. Cyclones are effective in removing the larger and heavier particles. For finer particles, electrostatic precipitators and bag filters achieve high separation efficiency. Energy consumption arising from fluid flow through equipment is again a source of utility waste.

### 5.6.3 *Separation of Liquid Particulates from Gas Streams*

Entrainment of liquid particulates in gas streams is common in a number of process operations. Examples include gas–liquid contacting devices used for gas absorption and distillation, vapour liquid separation vessels including steam drums, and heat exchangers where liquid is condensed from a gas stream during cooling. A distinction is sometimes made between *spray* (where the liquid particulate size is in the range 10 μm to >1000 μm) and *mist* (where the liquid particulate size is in the range (10 μm to 0.01 μm).

The finer the liquid particles, the more difficult the separation, and the more complex and costly the separation equipment required.

Mists are usually formed in three ways:

- mechanical shearing of films, droplets or bubbles
- cooling and subsequent condensation of liquid particles from a gas
- chemical reaction of two or more gases to form a product with a relatively low vapour pressure which condenses from the gas.

The mechanism of mist formation and selection of mist elimination equipment is discussed by Ziebold (2000). Energy consumed by the passage of gas through the mist eliminator is again a source of utility waste.

## 5.7 Drying

Wet solid materials are dried in a range of equipment, often involving transfer of moisture from the wet solid to gas (commonly air). In such cases, characteristic sources of waste include

- solids loss through entrainment in gas
- fouling of heat transfer surfaces
- cleaning of deposits
- product degradation due to heat effects, particle degradation.

Energy consumption occurs both through heating and materials transport.

| Energy use | Utilities | Wastes |
|---|---|---|
| Heating | Steam or fuel | Utility wastes |
| Transport of gas, solids | Electricity | Utility wastes |

Liquids can also be converted to solids through spray drying which is widely employed in the dairy industry. The term drying has also been used to describe the removal of water from gases. This usually requires an agent material. An example of gas drying using an agent material is given in Section 5.11.

## 5.8 Evaporation and Condensation

Evaporation of aqueous solutions is used as a means of increasing the concentration of a component in a solution. Sometimes evaporation is used as a precursor to crystallisation and/or drying of a solid product. Where heat-sensitive liquids are involved, vacuum evaporation is often employed to lower process temperatures. Wastes generated in evaporation typically involve

- products of fouling or cleaning arising from deposits caused by concentration and temperature effects
- concentration of non-volatile impurities in the evaporated liquid, and any associated purges
- product loss through entrainment of liquid in vapour.

There are several sources of energy required.

| Energy use | Utilities | Wastes |
|---|---|---|
| Heating medium | Steam | Utility wastes |
| Liquid circulation | Electricity | Utility wastes |
| Vapour recompression | Electricity | Utility wastes |
| Vacuum generation | Electricity or steam | Utility wastes |

Opportunities for energy conservation exist through using evaporated steam as a heating medium for subsequent stages of evaporation. Evaporated steam is sometimes compressed to achieve higher steam temperatures for heating in subsequent stages of evaporation.

Condensation by cooling can also be used as a means of partially separating components of different vapour pressure from a vapour stream. Residual separation tasks may still result if the condensate contains multiple components, or if further removal of a component from the vapour stream is required. Condensation is an important adjunct to other separation processes such as distillation and adsorption.

## 5.9 Solid–Liquid Extraction

Solid–liquid extraction involves the separation of desired solids from solid residues using a suitable solvent. Thus hot water is used with coffee grounds to extract coffee, while acids or alkalis are used to leach valuable ores from mined materials. In ore leaching, agitated vessels are commonly used to achieve solid liquid contact; heat is sometimes used to speed the rate of solid dissolution. Common sources of waste in solid–liquid extraction include

- disposal of exhausted solids
- leakage, spillage, purge of solvent
- corrosion products, for example, in acid leaching of minerals.

There are several sources of energy required.

| Energy use | Utilities | Wastes |
|---|---|---|
| Liquid pumping | Electricity | Utility wastes |
| Agitation | Electricity | Utility wastes |
| Heat (in some cases) | Steam | Utility wastes |

## 5.10 Liquid–Liquid Extraction

Perforated plate columns, centrifugal extractors or mixer-settler units are commonly used to enable transfer of a solute from one immiscible liquid phase to another. Characteristic wastes generated include

- contaminants concentrating at the liquid–liquid interface
- liquid losses, spillages
- contamination of solvent
- contamination of aqueous product streams.

There are two sources of energy required.

| Energy use | Utilities | Wastes |
|---|---|---|
| Liquid pumping | Electricity | Utility wastes |
| Agitation | Electricity | Utility wastes |

## 5.11 Use of Extraneous Materials

As observed for reactors in Chapter 4, the use of extraneous materials in separation processes can also lead to product contamination and waste generation. Waste generation applies particularly to spent extraneous material. Extraneous materials have been used industrially because they offer specific technical and economic advantages. However, the environmental consequences of their use are not always fully recognised or evaluated. Some examples of using extraneous materials in separation processes include

- activated carbon, diatomaceous earth in filtration
- chemicals in flocculation, for example, $AlCl_3$, $FeCl_3$
- poly-electrolytes in sedimentation
- agents used in gas drying.

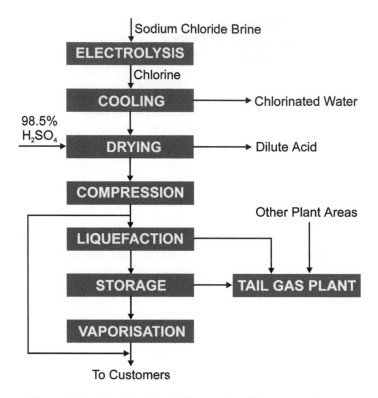

**Figure 5.4**   Simplified block diagram for chlorine purification.

### 5.11.1 *Example of Extraneous Material Use — Sulphuric Acid in Chlorine Drying*

Chlorine gas leaves electrolytic cells typically at around 90°C, 120 kPa abs, and saturated with water vapour. Before being stored or transported, chlorine must be thoroughly dried to avoid corrosion in steel equipment and piping. A simple block diagram for a commonly used approach to chlorine drying, compression and liquefaction is shown in Fig. 5.4.

Sulphuric acid (98% w/w) is an effective drying agent for gases, and is widely used for drying chlorine. The resulting dilute acid is often a waste product from chlorine plants. Reconcentration of the waste dilute acid is technically feasible but requires high temperatures and vacuum, and becomes increasingly difficult when evaporating to higher acid concentrations. Acid consumption can be minimised by use of refrigeration in both the cooling and the drying steps, subject to avoiding solid chlorine hydrates, which form around 14°C or lower. The problem of waste sulphuric acid in chlorine drying is further addressed in Problem 2, Section D.

## 5.12 Case Examples

Design and operation of both process and equipment can contribute to waste formation in separation processes. We first consider a case related to column internals selection in gas absorption equipment. Tray or packed columns are commonly used for gas absorption. Figure 5.5 shows typical internals for a packed column.

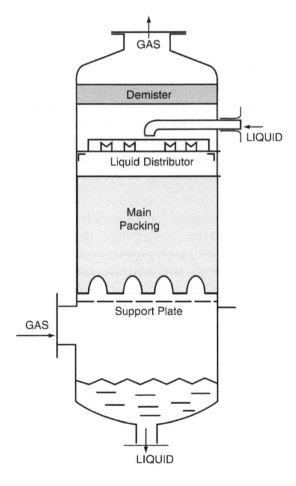

**Figure 5.5** Packed tower and internals used for gas–liquid contacting.

### 5.12.1 *Case Example — Solid Sodium Cyanide Plant*

Internals selection in gas–liquid contacting columns (Mudaliar and Sparrow, 2001)

Figure 4.4 showed a simplified flow sheet for the manufacture of HCN, a precursor for sodium cyanide manufacture. A sodium cyanide plant at Yarwun, Queensland, Australia has operated a carbon dioxide removal unit to remove $CO_2$ from the natural gas feed to the plant. The presence of $CO_2$ in gas leaving the reactor has two downstream effects, both undesirable:

- Caustic soda is consumed in the HCN absorption step and reacts with $CO_2$ to produce $Na_2CO_3$.
- $Na_2CO_3$ is an impurity in the NaCN product, and causes problems in conversion of the liquid NaCN to solid NaCN product.

In the Yarwun plant, $CO_2$ has been removed in an absorption column using Benfield solution (aqueous potassium carbonate/bicarbonate, vanadium, and methylamine). The Benfield solution is subsequently stripped of $CO_2$ in a 'regenerator' or stripping column before its re-use in the absorption column. Problems on the Yarwun plant had been encountered with the $CO_2$ absorption-stripping system. The absorber and stripper columns are essentially packed columns with some wash trays at the top for cleaning gas prior to leaving the columns.

**Plant investigations**

Investigation of the plant identified some key difficulties. Although $CO_2$ was being successfully removed from the gas feed, the gas throughput of the absorber was restricted, limiting plant production capacity. Further, Benfield solution entrainment in the gas was excessive, contaminating the catalyst bed in the downstream HCN reactor. The catalyst contamination led to frequent catalyst changes; this meant considerable plant downtime with loss of production capacity, and high costs incurred in replacing the expensive catalyst.

**Design changes**

These problems were addressed and overcome by some important design changes, especially in column internals as outlined below.

1. Wash trays in both absorber and stripper columns had been valve trays. Wash liquor rates across the trays were very low, leading to some weeping occurring. The trays were effectively operating under 'dry flooding' conditions, with the upward flow of gas entraining liquid as small droplets.

*Valve trays were replaced by bubble cap trays* which proved leak tight, and offered greater flexibility in terms of open area for gas flow. The caps also provided some inherent capability in intercepting entrained liquid. A *'picket and fence' design outlet weir* (Fig. 5.1) suitable for very low liquid rates, was used to replace conventional weirs on the tray.

2. A *mist eliminator mesh pad was installed* above the top tray in each column. For the absorber, less liquid was carried over to the downstream plant. For the stripper, the overhead vapour stream was cleaner, leading to cleaner recycle wash condensate.

3. *Bed limiters were installed* above the packing in both columns to prevent carryover of packing during surges of high gas flow.

4. The liquid distributors in the absorption column were found to have insufficient gas riser capacity and insufficient liquid distribution points. *New liquid distributors were installed providing an increase in gas riser area from 20% to 30% of column area and an increase in the number of liquid distribution points* from 100 to 150 drip points per square metre of tower cross-sectional area.

5. Additional liquid removal capability was provided downstream of the absorber to remove any residual liquid in the gas leaving the absorber. This comprised a new high performance cyclone and other modifications.

**Lessons**

This case highlights the importance of design and selection of internals for vapour–liquid contacting towers, and some potential effects of internals on performance of absorption and distillation systems. While plant capacity limitations and reliability were of prime concern in this case, the effects on waste minimisation are also evident, and could be of greater concern in other absorption and distillation systems. More details of the case including a process flow diagram for the absorber/stripper unit are provided by Mudaliar and Sparrow (2001). Details of column internals for gas–liquid contacting are found in Sinnott and Towler (2009) and could be obtained from equipment suppliers.

## 5.12.2 *Other Case Examples of Gas Absorption in Chemical Processes*

The removal of carbon dioxide from natural gas using gas absorption is widely practised in purification plants for wellhead natural gas. Carbon dioxide removal from flue gas by gas absorption is also part of many design schemes for carbon capture from power plants derived from fossil fuels. This case is addressed within Problem 3 in Section D. Gas absorption also features in common chemical processes, for example,

  (i)  absorption of $SO_3$ in 98% sulphuric acid in sulphuric acid manufacture
 (ii)  absorption of water in 98% sulphuric acid solution in gas drying in sulphuric acid manufacture
(iii)  absorption of formaldehyde in water in formaldehyde manufacture
(iv)  absorption of HCN in aqueous NaCN manufacture
 (v)  absorption of $NO_2$ in water in nitric acid manufacture.

It is important to note the process stream configuration around the absorption column in each application. Figure 5.2 showed a typical case of a gas absorption column arrangement. The solvent leaving the column base is recirculated to the top of the packing with both addition of the solvent and removal of product in the solvent recirculation loop. Where absorption is exothermic, heat is usually removed in an external heat exchanger. Consideration should be given in design and operation to the consequences of solvent loss, as well as incomplete removal of the absorbed species from the gas feed. Incomplete absorption is one aspect of Problem 3 in Section B, on waste minimisation in nitric acid manufacture.

### 5.12.3 *Case Examples in Distillation*

Distillation operations in industry are common in petroleum refining, petrochemical, and organic chemical process operations. Often the separations are complex and involve trade-offs between capital cost and separation efficiency. Ethylene manufacture (Fig. 4.3) involves a sequence of separations by distillation under pressure and refrigeration conditions. One key separation is that of ethylene product from unreacted ethane, which is recycled to the steam cracking reactor to supplement fresh feedstock.

A further case example occurs in *ethanol distillation* within the process for producing *fuel ethanol* from sugar based feedstocks. Fuel ethanol is commonly blended as a 10% by volume component with gasoline for the automobile fuel market. The sugar based feedstock is converted to ethanol by fermentation, but the stream leaving the fermenter contains typically less than 10% w/w $C_2H_5OH$ and may be considerably lower, at around 4% to 6% w/w $C_2H_5OH$. Much of the water present can be separated by distillation but an azeotrope in the $C_2H_5OH–H_2O$ system forms at around 90% w/w $C_2H_5OH$. Stringent water specifications (typically 0.5% by weight) are applied to fuel ethanol requiring an essentially anhydrous product.

Approaches to converting the azeotrope to anhydrous ethanol have historically involved azeotropic distillation, requiring an entrainer, which functions as an agent material. Examples of entrainers which have been used commercially include benzene and more recently cyclohexane. Benzene was replaced because of its toxic properties. Current approaches favour water removal by adsorption (discussed in Section 5.5) using molecular sieves.

Impurities in the fermenter effluent (comprising mainly ethanol and water) which is fed to the distillation step include

- fusel oils (heavier alcohols formed during fermentation), which must be removed as a purge from around the middle of the column
- non-condensable or volatile components, for example, carbon dioxide and aldehydes, which are vented from the condenser
- non-volatile components leading to a bottoms effluent stream which requires treatment for markets or disposal.

This case is of interest from a number of perspectives, including the

- replacement of benzene by cyclohexane, an inherently less hazardous entrainer from a safety and environmental viewpoint
- evolution of technology selection from azeotropic distillation to adsorption using molecular sieves
- use of agent materials (entrainers or molecular sieves) and their impacts as sources of waste
- range of impurities in the feed to the distillation step and their contributions to process waste
- combined energy required for the distillation and dehydration steps and its magnitude in relation to the energy value of ethanol.

## 5.13 Concluding Remarks

The approach adopted in this chapter serves as a useful guide and checklist for identifying waste sources and minimising waste in specific separation processes. The approach can be extended in detail for those separation processes discussed. The approach can also be applied to separation processes not discussed in this chapter.

Distinct sources of process waste apply for different separation processes. The quantity and composition of wastes often reflect the stringency and complexity of the separation duty. However, we can make some generalisations regarding sources of waste in separation processes.

Wastes occur as a result of

- imperfect separations, reflecting a range of technical and economic constraints;
- use of agent or extraneous materials such as solvents, filter aids, adsorbents, or desiccants; such materials ultimately require regeneration, treatment or disposal involving a range of environmental impacts;
- energy consumption, including hot and cold utility use, and through energy consumed in fluid flow through equipment or as a consequence of processing at elevated pressures.

## References

Mudaliar, G. and Sparrow, M. (2001). *Modification to CO$_2$ Removal Unit Prevents Cyanide Catalyst Poisoning*, Proceedings of World Congress of Chemical Engineering, Melbourne.

Sinnott, R. and Towler, G. (2009) *Chemical Engineering Design,* Butterworth-Heinemann, Oxford, UK.

Smith, R. (2005) *Chemical Process Design and Integration,* John Wiley and Sons, Chichester, UK.

Ziebold, S. A. (2000) Demystifying mist eliminator selection, *Chem. Eng.,* Vol. 107, May, 94–102.

## Chapter 6

# Identification of Waste in Utility Systems

## 6.1 Introduction

Process plants need utilities primarily as energy sources. Energy is needed for

- *heating duties* including steam, hot oil, other heat transfer fluids, and direct and indirect heating from hot gases
- *cooling duties* including cooling water, refrigeration
- *electricity* for drives on materials transport equipment (pumps, compressors, solid conveyors), agitators and centrifuges as well as plant support facilities such as lighting.

Other utility needs include

- nitrogen for purging
- compressed air for a variety of duties
- water for cleaning equipment and plant, and for fire fighting
- vacuum.

Environmental impacts result from the generation, distribution, and consumption of utilities. If a process plant uses a utility, it is indirectly responsible for impacts incurred in the generation and supply of the utility, including upstream impacts in the utility life cycle. Main impacts incurred in utility generation are consumption of fuel and water, and emission of fuel combustion products and aqueous effluents. Fuel combustion products

*Sustainable Process Engineering: Concepts, Strategies, Evaluation, and Implementation*
David Brennan
Copyright © 2013 Pan Stanford Publishing Pte. Ltd.
ISBN 978-981-4316-78-1 (Hardcover), 978-981-4364-22-5 (eBook)
www.panstanford.com

are predominantly gaseous, but coal combustion, for example, generates large quantities of solid waste. Since utilities are consumed at most stages of a product's life cycle, the aggregative effect from utilities in a product's environmental footprint is often substantial. Environmental impacts derived from utility use can be minimised *both* by

- reducing consumption of utilities on process plants and
- improving the design of the individual utility systems.

Figure 2.4 showed some examples of remote generation and supply of coal fired electricity, and of natural gas to a process plant. Environmental impacts in these cases result from

- mining, cleaning, and transport of coal supplied to the power station, electricity generation, and subsequent electricity transmission
- natural gas extraction, purification, compression, and pipeline transmission.

Figure 6.1 shows some of the interlinking of fuel and water resources in the generation of hot and cold utilities and electricity. Electricity generated from fossil fuels draws on fuel, water, steam, and cooling water. Electricity in turn is needed to drive boiler feedwater pumps in steam circuits, and cooling water pumps and air fans in cooling water systems. Electricity is needed for air blowers in fuel combustion systems. Refrigeration systems need electricity for gas compressors and cooling water for condensers. We now explore utilities under the separate headings of fuels, steam, cooling utilities, electricity, and selected other utilities.

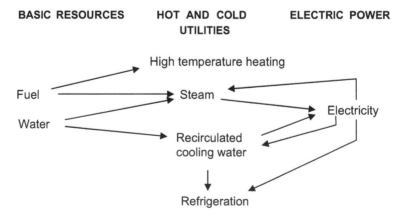

**Figure 6.1** Some key linkages between fuel, water, and common utilities.

## 6.2 Fuels

Fuel combustion provides a key source of energy for domestic, commercial, and industrial use. Important fuels used in transport, power generation, heating, and industrial processing include

- natural gas, primarily methane in composition
- liquefied petroleum gas (LPG), primarily liquid propane and butane
- crude oil processed in refineries to mainly liquid transport fuels
- coal, broadly classified as
  - ► lignite or brown coal
  - ► black coal
- bio fuels, for example, bio gas, fuel ethanol, biodiesel.

Transport, power generation, and gas supply are important to the wider community as direct users. Transport of raw materials and products are an integral part of industry structure and operations. Fuels are used by the process industries primarily for

- high temperature heating of process fluids
- steam and electricity generation.

Important drivers for fuel use in industry, commerce, and community include

- supply issues in terms of reliability, quality, and cost
- $CO_2$ and other emissions in fuel combustion
- resource depletion issues for fossil fuels
- possible use of renewable fuels or renewable energy sources.

## 6.3 Fuel Combustion

Furnaces, incinerators, gas turbines, diesel engines, flare stacks, and transport vehicle engines all produce waste through combustion products. The composition and quantity of waste from fuel combustion depends on the composition and state of the fuel and its mode of combustion. Fuel combustion wastes commonly include

- $SO_x$, mainly $SO_2$ but occasionally also some $SO_3$
- $NO_x$, mainly as NO and $NO_2$
- $CO_2$, CO
- unburned hydrocarbons
- particulates (entrained in flue gas)
- ash.

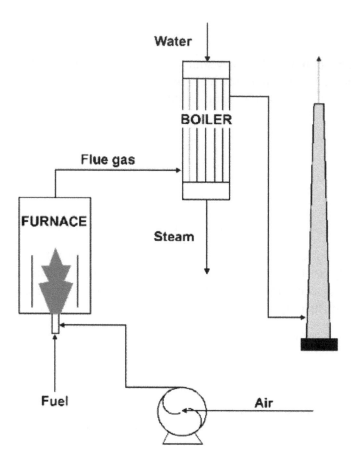

**Figure 6.2** Simplified diagram of fuel combustion system with steam generation.

Fuel is combusted in a furnace with excess air. Fans are required for the combustion air (and in some cases the flue gas) to overcome pressure losses in the gas stream. In the simplified diagram in Fig. 6.2, heat is recovered from flue gas to generate steam prior to discharge via a stack to atmosphere. In some cases particulate removal would occur before or after the boiler.

### 6.3.1 *Heat of Combustion*

Two methods are commonly used for reporting heating values (or calorific values) of fuels. The total heating value (or 'higher' or 'gross' heating value) is the heat evolved in the complete combustion of the fuel under constant pressure at 25°C when all water initially present in the fuel and in the combustion products is condensed to the liquid state. The 'net' or 'lower' heating value (or calorific value) is similarly defined, except that the final state of the water after combustion is taken as *water vapour* at 25°C.

## 6.3.2 *Excess Air*

Fuel is combusted with air (or oxygen, or oxygen enriched air) in excess of that required stoichiometrically to ensure complete combustion of the fuel. The presence of carbon monoxide in flue gas indicates that not all the energy potential from the fuel combustion has been achieved; further, carbon monoxide is toxic. Excess air ensures all carbon in the fuel is oxidised to $CO_2$ and all sulphur is oxidised to sulphur dioxide. The amount of excess air required for complete combustion is greater for solid fuels than gaseous fuels. Excess air requirements also depend on burner and furnace design. Excess air is typically 25% for coal combustion, and 10%–15% for gas combustion. A rule of thumb for excess oxygen content in flue gas is around 3% by volume.

## 6.4 Common Fuels

'Wellhead' gas from gas fields is processed to produce a 'sales' gas of defined properties and composition for reticulation and supply to industrial, commercial, and domestic consumers. Sales gas composition may vary from one region to another. Sales gas composition in Victoria, Australia is shown in Table 6.1. Wellhead gas composition from different gas fields may vary considerably (see Table 6.2 for examples). Wellhead gas may be produced from dry wells to produce predominantly gas. Often wellhead gas is found

Table 6.1   Sales gas composition in Victoria, Australia

| Component | % Vol | mg/Nm$^3$ | MJ/kg |
|---|---|---|---|
| $CH_4$ | 90.6 | | |
| $C_2H_6$ | 5.6 | | |
| $C_3H_8$ | 0.8 | | |
| $C_4H_{10}$ | 0.2 | | |
| $N_2$ | 1.1 | | |
| $CO_2$ | 1.7 | | |
| Sulphur | | 45 | |
| Water | | 60 | |
| Lower CV (calorific value) | | | 46.7 |

Table 6.2    Examples of variation in natural gas wellhead compositions

| Location | Ballera | Timor sea | Bass strait |
|---|---|---|---|
| Purification plant site | Southwest Queensland | Near Darwin | Victoria |
| Pressure (kPag) | 8,000 | 31,000 | 7,000 |
| Temperature (°C) | 60 | 135 | 23 |
| Molar composition (%) | | | |
| Methane | 61.6 | 70.7 | 80.2 |
| Ethane | 10.0 | 7.4 | 7.0 |
| Propane | 3.7 | 4.6 | 4.4 |
| i-Butane | 0.6 | 1.2 | 0.9 |
| n-Butane | 1.1 | 1.7 | 1.2 |
| i-Pentane | 0.3 | 0.8 | 0.6 |
| n-Pentane | 0.4 | 0.6 | 0.6 |
| $C^{6+}$ | 2.6 | 4.1 | 2.3 |
| Nitrogen | 1.6 | 3.4 | 0.7 |
| Carbon dioxide | 17.9 | 5.5 | 2.1 |
| Total | 100 | 100 | 100 |
| Water | saturated | saturated | saturated |
| Hydrogen sulphide | 30 ppm | 20 ppm | 60 ppm |

**Note:** Wellhead gases may also contain entrained water, mercury, and higher concentrations of sulphur.

in conjunction with liquid hydrocarbons which are separated from the gas to form 'condensate' (mainly liquid propane and butane) and 'crude' oil. Wellhead gas must be purified to remove liquid hydrocarbons, water, sulphur, and $CO_2$ to meet sales gas specifications.

Crude oil comprises higher molecular weight hydrocarbons and typically contains 84%–87% carbon, 11%–14% hydrogen, 0%–3% sulphur, and 0%–0.6% nitrogen by weight. Wellhead crude oil from predominantly oil wells is 'stabilised' to separate dissolved or entrained hydrocarbon gases. Crude oil is converted into a wide range of (mainly) liquid fuels within a petroleum refinery. The initial processing step in the refinery is crude oil distillation, which separates oil into various boiling point fractions. These fractions are then further processed in the refinery (an integrated complex of plants), producing a range of fuels such as LPG, petrol, diesel, aviation fuel, kerosene, and fuel oil. On distillation of the crude oil in the crude distillation unit (CDU), sulphur becomes more concentrated in the heavy fuel fractions. Increasing effort has focused in recent years on improving the quality of transport fuels such as petrol and diesel with respect to reducing the quantity of

- sulphur (diesel and gasoline)
- benzene (gasoline).

Table 6.3   Examples of coal compositions

|  | US black coal (mass percent)[1] | Victorian brown coal (mass percent) |
|---|---|---|
| Moisture | 2.7 | 62 |
| Carbon | 73.6 | 25.7 |
| Hydrogen | 4.6 | 1.8 |
| Sulphur | 2.5 | 0.11 |
| Nitrogen | 1.3 | 0.23 |
| Ash | 8.4 | 0.84 |
| Oxygen | 6.9 | 9.4 |
| Total | 100 | 100 |

**Note:** [1]Cohen Hubal and Overcash (1993).

Effort has also focussed on combustion engines with respect to achieving complete fuel combustion and improved engine efficiency.

*Coal* analyses are sometimes reported as *ultimate* (each major chemical element), and sometimes as *proximate* (moisture, ash, volatile matter, and fixed carbon). The classification of coal depends largely on calorific value, carbon content, and water content. Lignite and sub-bituminous coals are of lower grades, with higher concentrations of water. Over extended geological time periods, with exposure to higher temperatures and pressures, these coals become naturally harder forming bituminous coals and ultimately anthracite. Coal compositions vary widely from country to country, and within countries. Some examples of coal compositions are provided in Table 6.3.

*Fuel ratio* of coal is defined as the ratio of its percentage of fixed carbon to that of volatile matter. The *rank* of the coal may be estimated from its fuel ratio.

| Rank | Fuel ratio |
|---|---|
| Anthracite | 10–60 |
| Semi-anthracite | 6–10 |
| Bituminous | 3–7 |
| Semi-bituminous | 0.5–3 |

## 6.5  Environmental Impacts of Flue Gases

Table 6.4 shows some of the more common emissions from combustion of fuels. Resulting environmental impacts, discussed in more detail in Chapter 10, include

- acidification ($NO_x$ and $SO_x$)
- global warming ($CO_2$)

Table 6.4 Common emissions from combustion of fuels

| Component in fuel | Fuel type (coal, fuel oil, natural gas, transport fuel) | Product or residue from combustion |
|---|---|---|
| Sulphur | From fuel | $SO_2$ |
| Nitrogen | From fuel | $NO_x$ (fuel) |
| | From combustion air | $NO_x$ (thermal) |
| Carbon | From fuel | $CO_2$ (complete combustion) |
| | | CO (incomplete combustion) |
| Hydrogen | From fuel | $H_2O$ |
| Ash* | Coals (mainly) | Ash residue particulates in flue gas |
| Heavy metals, including mercury | Coals (mainly) | Heavy metals to air or ash |
| Mercury | Some oil and gas sources | Normally captured prior to combustion |

**Note:** *Ash* is the inorganic residue resulting from burning of coal. It consists mainly of $SiO_2$, $Al_2O_3$, $Fe_2O_3$, CaO with smaller amounts of MgO, $TiO_2$, alkali, and sulphur compounds.

- nutrification ($NO_x$)
- smog from photochemical oxidation ($NO_x$, unburned hydrocarbons)
- fuel resource depletion
- human health effects derived from $NO_x$, $SO_2$, CO, particulates, traces of heavy metals.

These impacts may be reduced by minimising utility of fuel consumption but also by

- switching to a cleaner fuel
- switching to a fuel of higher calorific value
- improved burner design
- treating fuel before combustion to minimise impurities
- treating stack gas after combustion to remove certain pollutants, for example $SO_2$
- improving control and/or operation of the combustion process.

## 6.5.1 *$NO_x$ Formation in Fuel Combustion*

When the temperature of combustion is high enough (approximately >1000 K), some of the nitrogen and oxygen from the combustion air react to form nitrogen oxides, principally NO and $NO_2$, represented as $NO_x$. Since the extent of this reaction depends on temperature, the $NO_x$ thus formed is termed *thermal $NO_x$*.

Many fuels, for example, coal, contain nitrogen which can react with oxygen in the combustion air during combustion to form $NO_x$. Some coals have as much as 3% wt N. The $NO_x$ thus formed is termed *fuel* $NO_x$. Most $NO_x$ emissions are in the form of NO which subsequently oxidises to $NO_2$. On cooling, $NO_2$ becomes $N_2O_4$ which is brown in colour.

$NO_x$ emissions in combustion systems can be reduced by

- lowering the temperature of fuel combustion (e.g., by using catalytic combustion)
- using fuels which are lower in nitrogen
- using low $NO_x$ burners which employ a staged combustion process
- using less excess air
- injecting ammonia and catalytically reducing $NO_x$ to nitrogen and water.

These options are discussed in more detail by Smith (2005).

## 6.6 Theoretical Flame Temperatures

Theoretical flame temperatures (also called 'adiabatic flame temperatures') can be calculated for the combustion of various fuels. The calculations assume rapid, adiabatic combustion, where the energy released by combustion is taken up by gases leaving the combustion zone (Felder & Rousseau, 2005, Section 9.6; Smith, 2005, Section 15.10.). The specific heats for components of air and combustion product gases are temperature dependent — this must be allowed for in calculations. In practice, heat losses from the flame and dissociation reactions cause the flame temperature to be less than the theoretical or adiabatic flame temperature. Some theoretical flame temperatures for stoichiometric mixtures with air are as follows:

methane 1875°C
propane 1925°C
hydrogen 2045°C.

## 6.7 Furnaces

High temperatures required for process streams and reactors are achieved in furnaces. Furnace efficiency is limited by heat lost in flue gas discharged to atmosphere. Furnace efficiency can be expressed as the ratio

$$\frac{\text{heat absorbed by process}}{\text{heat released by fuel combustion}}$$

Furnaces are usually operated with the flue gas exhausted to atmosphere above the dew point of the flue gas. The dew point is the temperature at which the onset of condensation occurs during cooling of the flue gas. Condensate can be corrosive to equipment and piping due to low pH derived from dissolved acidic gases, and also due to nitric and sulphuric acids resulting from $NO_x$ and $SO_x$ in the flue gas. If there is sulphur in the fuel, the stack temperature is kept above 160°C; fuels free of sulphur can be cooled to lower stack temperatures.

Refractory materials are used to line the inside of furnaces. Important properties apart from insulating capability include resistance to attack by the process environment, and also strength properties such as refractoriness, re-fractoriness under load, thermal spalling resistance, and cold (compressive) strength.

The design of fuel burners is important in ensuring complete and efficient combustion with minimal pollutants. The design is distinct for solid, liquid, and gaseous fuel burners. Perry's Chemical Engineers Handbook provides diagrams of various types of industrial burners (Perry & Green, 1984, Figures 9.32 to 9.60). Solid fuels such as coal use grates for small capacity and pulverised fuel for large capacity. Liquid fuels are atomised and vaporised into the combustion air stream. Gaseous fuels are mixed with combustion air in a variety of burner designs. The importance of the three T's in burner design is frequently emphasised—temperature, turbulence, and (residence) time. Some discussion of fired heaters with diagrammatic layouts is provided in Sinnott and Towler (2009, Section 12.17). Fired heaters are used in industry for combusting fuel to provide heat to process fluids where required process temperatures exceed those achievable by steam heating.

## 6.8 Flare Stacks

Flare stacks are used in gas plants, petroleum refineries, and petrochemical plants to combust surplus hydrocarbons to produce combustion products that are neither toxic nor combustible. Flares frequently incorporate a liquid–gas separator at the base of the stack and steam assisted burner nozzles at the top of the stack to aid complete combustion. The diameter of the flare must facilitate a stable flame and prevent 'blowout' (where vapour velocities exceed 20% of sonic velocity). The height of the flare must be adequate to prevent radiation damage to humans and equipment (and other plant hardware). A common design criterion is to ensure heat intensity at the base of the stack is less than 4.5 kW/m$^2$ (corresponding to the pain threshold for exposed human skin).

## 6.9 Steam Generation

Steam is generated in a *boiler*. There are two types of boiler:

- *Fire tube,* where hot gases flow through tubes surrounded by a pool of boiling water. Steam pressure is generally restricted to around 20 bar g. It is normally used for smaller steam generation rates and lower steam pressures.
- *Water tube,* where water is circulated by natural convection between a water drum and steam drum interconnected by tubes (boiling water rises, saturated water descends). Hot gases flow through an external chamber.

*Superheaters* are heat exchangers used to superheat saturated steam. They may be integral with water tube boilers or may be quite separate heat exchangers.

*Economisers* are heat exchangers (usually separate from the boiler) used to heat water to boiling temperature corresponding to the steam generation pressure.

*Typical boiler pressures* are

- 100 bar g for power generation
- 10–40 bar g for distribution within a process plant for heating (a minimum pressure for distribution would normally be 2–5 bar g).

Higher pressure boilers incur higher capital costs because of greater demands on boiler material properties and greater wall thickness, and on boiler construction standards.

A simplified flow sheet for steam generation is shown in Fig. 6.3. Feedwater (typically towns water) is normally demineralised, for example, by ion exchange, and mixed with recovered condensate from the steam circuit. The returned condensate may in some cases need treatment, for example, if there is risk of contamination by a process stream. The mixed stream of boiler feedwater is fed to a deaerator, where dissolved oxygen is removed by sparging the water with steam at around 100°C; oxygen solubility is reduced by the increased temperature. A small amount of steam containing the non-condensable gases is vented to atmosphere. Some residual oxygen remains; this is then removed chemically, for example, by sodium sulphite

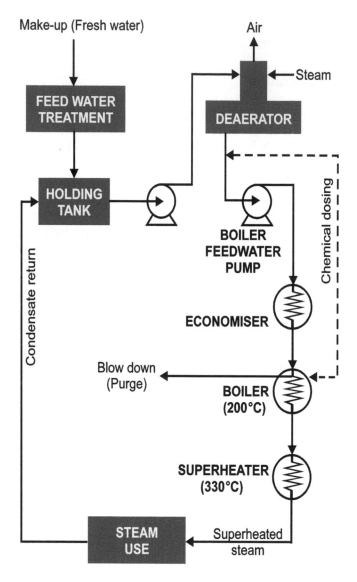

**Figure 6.3** Water circuit for steam generation.

$(Na_2SO_3 + 0.5O_2 \rightarrow Na_2SO_4)$ or hydrazine $(N_2H_4 + O_2 \rightarrow 2H_2O + N_2)$ or other organic compounds. Certain dissolved solids remaining in the water after demineralisation (e.g., calcium and magnesium) can be precipitated by sodium phosphate addition.

Table 6.5    Common standards for boiler water quality

| Boiler drum water quality for 40 bar g boiler | ASME boiler guidelines | British Standards (BS 2486) |
|---|---|---|
| Maximum total dissolved solids | 500–2500 mg/L | 1000 mg/L |
| Maximum alkalinity | 250 mg/L | 25–50 mg/L |
| Maximum total suspended solids | 8 mg/L | Not specified |
| Maximum conductivity | 800–3800 μS/cm | 2000 μS/cm |
| Maximum silica | 40 mg/L | 20 mg/L |

**Note:** S denotes siemens, the unit of conductance $=$ A (ampere)/V (volts) $= \Omega$ (ohm)$^{-1}$. Most drinking water sources have a conductivity in the range 500 to 800 μS/cm.

Demands on *boiler water quality* become more exacting as boiler pressure (and hence temperature) increases. Key considerations are total dissolved solids and oxygen concentration, which influence corrosion in the boiler. Steam quality (influenced by entrainment of boiler water) is more demanding for steam turbine applications than process heating applications. Silica, which affects scaling on steam turbine blades, is removed by ion exchange in feedwater treatment. Some common industry standards for boiler water quality are ASME (American Society of Mechanical Engineers) Boiler Guidelines and the British Standard (BS 2486:1997). Excerpts from these standards are shown in Table 6.5 for a 40 bar g boiler.

Any suspended solids or dissolved solids in the boiler water are concentrated by evaporation during boiling. Therefore water must be purged from the boiler to control the solids concentration; excessive concentrations cause corrosion and fouling. This purge is often referred to as 'blowdown' (reflecting the associated pressure reduction). The ratio of impurity concentration in the boiler drum to that in the boiler feedwater pumped from the deaerator is known as the number of 'cycles of concentration'.

Blowdown rate, usually expressed as a percentage of the evaporation rate, depends on the effectiveness of boiler feedwater treatment and boiler pressure. Typical blowdown rates are

- small, low pressure boilers: 5%–10%
- large, high pressure boilers: 2%–5%.

The most common heat source for steam generation is fuel combustion, most commonly natural gas, but coal, fuel oil and other fuels are also used. Natural gas boilers are cheaper in capital cost than coal fired boilers and have fewer overall emissions, but gas is a more expensive fuel than coal.

Important sources of heat in chemical processes are high temperature process streams, where the process stream is cooled in a 'waste heat boiler'. Waste heat boilers are integral with many process technologies such as sulphuric acid, nitric acid, hydrogen cyanide, formaldehyde, and ethylene manufacture, as discussed in Chapter 4. Processes involving reactions at elevated temperatures, especially exothermic reactions, lend themselves to heat recovery opportunities. Constraints often apply to exit gas temperatures from process gases arising from consideration of effective temperature driving forces, condensation of acids, and operating conditions in downstream process equipment.

## 6.10 Steam Use

Steam is used mainly for process heating or for driving turbines, but can be used as a raw material (e.g., in steam reforming of methane to produce hydrogen), as a diluent in a process (e.g. in steam cracking of hydrocarbons to produce ethylene), for steam stripping in a process, for cleaning plant and equipment, or for steam jet ejectors to produce vacuum. Examples of steam stripping are in VCM recovery from PVC manufacture (Chapter 4) and in solvent recovery processes by absorption stripping (Chapter 5).

As a heating medium, steam is condensed isothermally at its saturation temperature. Its condensation temperature, influencing the temperature driving force (or $\Delta T$) with respect to the process stream, and latent heat of condensation influencing steam consumption, are key properties. The higher the pressure of saturated steam, the higher the condensing temperature but the lower the latent heat in kJ/kg (refer to steam tables and thermodynamic property diagrams). There is an upper pressure limit for which saturated steam can be used (identify this on a pressure enthalpy diagram). Condensing steam has a high heat transfer coefficient, nominally $\sim 8000 W/m^2 K$, but this is reduced if small concentrations of non-condensing inert gases such as air are present.

When used to power turbines (e.g., for electricity generation or for driving compressors or pumps), steam expansion through the turbine approximates an isentropic expansion. Inlet steam must be superheated, while exhaust steam is close to saturation. Exhaust steam may be at low pressures (even at vacuum as in power stations to maximise energy recovery), or intermediate pressures allowing exhaust steam to be used for heating purposes.

Often a plant will use steam at multiple pressures — for example, high, medium, and low pressure. High pressure superheated steam will be used

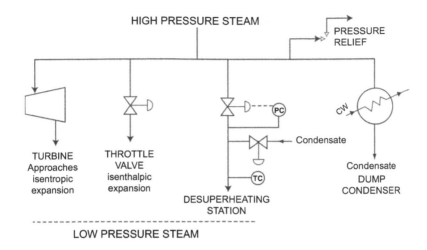

**Figure 6.4** Alternative uses of high pressure steam.

for driving machinery and lower pressure steam will be used for heating. Such systems can become complex to operate, as steam demands at the different pressures may vary with changes in plant operating conditions. There are advantages in distributing steam at higher pressures; steam mains are smaller in diameter and larger pressure drops can be accommodated. High pressure steam can be let down to lower pressures through throttling valves, but the expansion in this case is isenthalpic, with no useful work done.

Figure 6.4 illustrates some alternative uses for high pressure steam, including the isentropic expansion through a turbine and isenthalpic expansion through a throttling valve. If the exhaust temperature from the throttling valve is too high for some applications, desuperheating will be required. If demand for steam suddenly reduces below supply, pressure relief will be necessary resulting in loss of steam to atmosphere. This loss can be avoided by providing a dump condenser which allows condensate recovery.

Figure 6.5 shows an example of heat recovery in a hypothetical chemical process. Heat is recovered from the reactor to generate saturated steam in a boiler. The saturated steam is superheated using further heat recovered from the reactor in a superheater. The resulting superheated steam is expanded in a turbine to drive a centrifugal compressor in the process. Exhaust steam from the turbine is used both for live steam within the process, and as low pressure saturated steam in process heat exchangers. Condensed steam from the heat exchangers is collected as condensate, and demineralised water is added to the circuit. Condensate is pumped through a deaerator to the steam drum. The steam drum allows water to recirculate through the reactor waste

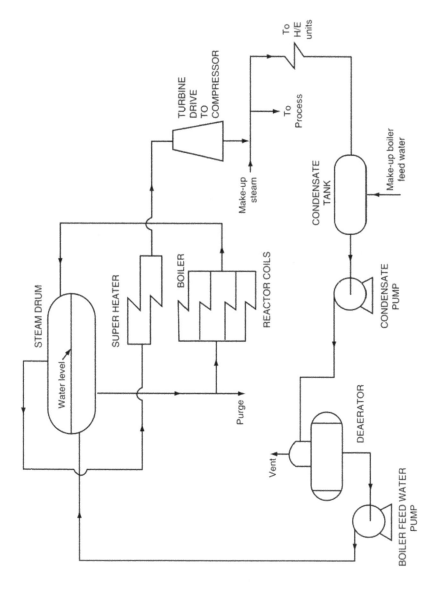

**Figure 6.5** Example of heat recovery from a chemical process.

heat boiler where it is partially vaporised, returned to the steam drum and separated into steam and condensate.

Condensate recovery is an important aspect of steam circuits. Condensate (if uncontaminated) is high quality hot water, usually pressurised, and valuable in energy content. Its recovery is important to minimise the quantity of fresh demineralised water make-up to the boiler circuit, and to maximise energy efficiency by burning less fuel. High pressure condensate may in some cases be usefully let down in pressure to generate lower pressure steam and condensate.

*Steam traps* must be placed in discharge lines from heat exchange spaces (whether heater, evaporator, dryer or other) to recover condensate efficiently.

The steam trap

- stops condensate accumulation which would reduce heat exchange efficiency and
- prevents steam escaping from condensate discharge lines.

Some traps allow removal of non-condensable gases. Many types of steam traps are used, for example, thermodynamic, mechanical, and orifice types. For details see http://www.spiraxsarco.com.

Insulation is important for equipment and piping carrying steam and condensate, both to minimise energy loss and provide personnel protection.

There are limits to the use of steam as a heating medium derived from its thermodynamic properties. Alternative heating media must be found for higher temperatures. Examples include Dowtherm (diphenyl/diphenyl oxide eutectic), molten salts (e.g., sodium and potassium salt eutectic mixtures) and mineral oils (hot oil). Fired heaters are widely used in these systems. Fired heaters (usually using natural gas) can also be used for heating process streams; fired heaters are common in petroleum refineries, for example, for heating heavier crude and residue oils.

## 6.11  Water Sources and Uses

*Water sources* include towns water for domestic use (e.g., from mains water supply), surface water (rivers, lakes, dams), ground (bore) water, recycled waste water, recirculated cooling water (from cooling towers), and sea water. Water quality varies considerably depending on its source and any upstream use. Table 6.6 gives examples of water compositions from one towns water

Table 6.6   Examples of raw water compositions

|  | Towns | Bore |
|---|---|---|
| Total hardness | 9 ppm | 1018 ppm |
| Sodium | 5 ppm | 527 ppm |
| pH | 6.7 | 7.8 |
| Free carbon dioxide | 4 ppm | 22 ppm |
| Total dissolved solids | 26 ppm | 2397 ppm |
| Sulphate | 1 ppm | 321 ppm |
| Chloride | 5 ppm | 860 ppm |
| Silica | 6 ppm | 32 ppm |
| Conductivity ($\mu$S/cm) at 20°C | 35 | 3300 |

**Note:** Suspended solids may also be present, especially in bore water.

and one bore water source. Water may require pretreatment before use in specific applications.

Water may be used for many applications in process plants, such as

- processing raw material either as *water*, for example, chloralkali from salt, or *steam*, for example, hydrogen from methane
- make-up to a cooling water circuit involving cooling towers
- make-up to a boiler feedwater circuit for steam generation
- firefighting
- cleaning
- drinking, washing, sanitary, garden.

### 6.11.1 *Water Quality Indicators*

Some commonly used water quality indicators are identified below.

*Hardness* reflecting the concentration of dissolved calcium and magnesium salts (<50 mg/L is classified as soft). Hardness is an indicator of scaling tendency leading to reduced heat transfer and increased pressure drop.

*Total dissolved solids*, normally inferred from conductivity measurement (1S corresponds approximately to 0.7 mg/L). Total dissolved solids is an indicator of corrosion and scaling tendency.

*Conductivity (electrical)*, measured as micro Siemens per cm., which is an indicator of corrosion potential.

*Alkali* delignifies timber which is often used for packing in cooling towers.

*pH* and *dissolved oxygen* are indicators of corrosion potential.

*Sulphate* attacks cement and concrete, often used in cooling tower basins.

*Chlorides* have implications for specific types of corrosion such as stress corrosion cracking and pitting of austenitic stainless steels.

## 6.12 Recirculated Cooling Water from Cooling Towers

Cooling water is commonly obtained from evaporative cooling towers (Fig. 6.6), where air flows upwards countercurrent to downward water flow. Area for combined heat and mass transfer in the cooling tower is provided by internal packing of timber or plastic. Electricity is consumed by air fans and cooling water recirculating pumps. Water losses occur in the cooling water circuit arising from

- entrainment of water droplets in air leaving the cooling tower, also termed 'drift'
- evaporation necessary to achieve cooling in the cooling tower
- any leakages in the cooling water distribution network
- purge necessary to maintain quality of the circulating cooling water.

These water losses must be replaced by fresh water make-up.

The purge is necessary to control the concentration of

- dissolved solids
- suspended solids.

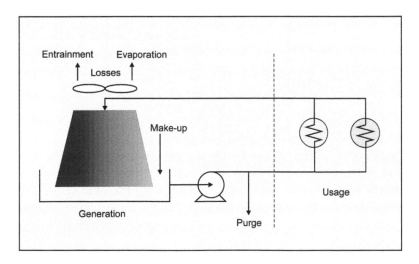

**Figure 6.6** Simplified flow sheet of recirculated cooling water circuit.

Concentrations of dissolved and suspended solids increase as a result of the evaporation process. Concentration of dissolved solids may be several multiples of that in the make-up water. Any one of several quality parameters might be controlling, for example,

- alkalinity (delignifies timber)
- hardness (influences scaling on heat exchanger surfaces)
- sulphate (influences attack on concrete basin of cooling tower).

Mass balances for overall water and component impurities can be made over the cooling water system to determine purge rates. For example, consider a system where evaporation $(E)$ and drift losses $(D)$ are estimated to be 3% and 0.2% respectively of the circulating water flow rate $(C)$. Assume make-up water contains 40 ppm dissolved solids while recirculated cooling water is to contain a maximum of 200 ppm dissolved solids. Note that in this case, dissolved solids is stipulated as the key impurity governing blowdown. The make-up water requirements $(M)$ and the blowdown (or purge) requirements $(B)$ can be estimated as a percentage of the circulating cooling water flow rate.

$$\text{Overall balance} \qquad M = B + E + D$$

Hence
$$M/C = B/C + E/C + D/C$$
$$= B/C + 0.032 \qquad (6.1)$$

$$\text{Dissolved solids balance} \qquad 40M = 200B + 200D + E \times 0$$

Hence
$$40M/C = 200B/C + 200D/C$$
$$= 200B/C + 200 \times 0.002$$
$$= 200B/C + 0.4 \qquad (6.2)$$

Solving Eqs. (6.1) and (6.2)

$$B/C = 0.0055 = 0.55\%$$
$$M/C = 0.0375 = 3.75\%$$

The extent of cooling achieved in the cooling tower is governed by the rate of water evaporation, which in turn is limited by the *wet bulb temperature* of ambient air. Cooling water temperatures leaving the base of the cooling tower will be lower in winter than summer. Cooling water temperatures also vary with locations — cooling water will be colder in northern European countries, than in Asian or Australian cities.

The design cooling water outlet temperature from heat exchangers is typically limited to 45–50°C because of scaling on heat exchanger surfaces.

This is a commonly adopted 'rule of thumb' in design practice. In some cases, a lower outlet temperature of cooling water will be selected in order to achieve target temperatures in the cooled process stream.

The potential for re-use of purged water from cooling tower circuits, or alternatively for treating aqueous effluents from other sources for make-up to cooling water circuits, should be examined.

Centrifugal pumps, typically of large capacity, are normally used for pumping cooling water through the plant and back to the top of the cooling tower. *Pump discharge pressure must be sufficient to*

- overcome pressure drops in piping and heat exchangers in the cooling water distribution system and
- reach required elevations, for example, distillation column condenser, top of cooling tower.

Very high reliability is required for cooling water supply in process plants necessitating multiple pumps, often involving a combination of electrical motor and steam turbine drives.

## 6.13 Sea Water Cooling

For plants located near the coast, sea water is an option for cooling because of its abundance and low first cost. Sea water contains about 3.4% salt (mainly NaCl, but also Ca and Mg salts) and is slightly alkaline. It is a good electrolyte and can cause galvanic corrosion, as well as pitting and stress corrosion cracking with stainless steels. Corrosion is influenced by oxygen content, velocity, temperature, and the presence of biological organisms (Fontana, 1986). The choice of sea water cooling will usually require more exotic materials of construction (and hence higher capital cost) for heat exchangers.

Return cooling water may cause environmental damage to the sea either because of thermal effects or because of contamination by impurities from leaking heat exchangers. The former risk will limit the return cooling water temperature, thereby increasing cooling water flow rate and associated pumping costs. The latter risk can be greatly minimised by using a closed loop secondary coolant, which in turn is cooled by sea water (Fig. 6.7). The closed loop coolant can be of such a quality that heat exchanger corrosion by the coolant is minimal. Since a temperature driving force $\Delta T$ exists between sea water and the closed loop coolant, and between the closed loop coolant and each process heat exchanger, there may be a penalty in terms of the lower limits of process temperatures achievable.

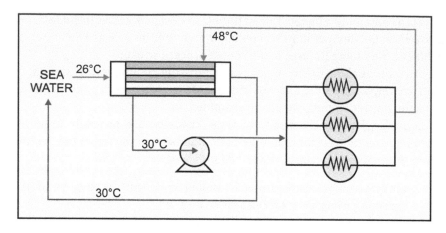

**Figure 6.7**  Example of sea water cooling with secondary cooling circuit.

An *important benefit* of sea water cooling is that it removes the demand for fresh water consumption as make-up to a cooling water circuit, important in locations where fresh water is a scarce resource.

## 6.14  Air Cooling

An alternative to water cooling is air cooling, using fans and extended surface heat exchangers. Extended surfaces are usually achieved by fins attached to the heat exchanger tubes. Electricity becomes the sole utility used. Extended surface is necessary because of the lower heat transfer coefficient of the gas compared to that for water. (Forced convection heat transfer coefficient for water is $\sim6000$ W/m$^2$K compared with $\sim200$ W/m$^2$K for air.) Some introductory notes on air cooled heat exchangers are provided in Section 12.16 of Sinnott and Towler (2009).

## 6.15  Refrigeration

Cooling of process streams using cooling water is limited by the cooling water temperature. If a design inlet cooling water temperature is limited to 28°C minimum, process stream temperatures will be limited to approximately 35°C, allowing for temperature rise in cooling water and temperature driving forces in a heat exchanger. Colder temperatures for process streams can be achieved using refrigerated coolants or evaporating refrigerants.

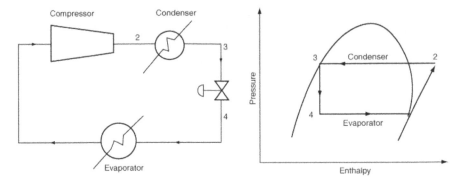

**Figure 6.8** Simple refrigeration circuit and corresponding cycle on pressure enthalpy diagram.

Refrigeration is part of many process industry plants. For example,

- in ethylene manufacture, it is necessary to separate ethylene the desired product from other hydrocarbons. This is done by distillation at low temperature ($-100°C$), requiring refrigeration, usually employing ethylene and propylene as refrigerants.
- in polyvinyl chloride (PVC) manufacture from vinyl chloride monomer (VCM), it is necessary to recover VCM from vent gases containing VCM and inert gases; refrigeration is used to condense and recover the VCM from the gases prior to incineration of the residual VCM.
- in chlorine liquefaction within a chlor alkali plant, refrigeration is required to liquefy chlorine gas.

A basic refrigeration circuit (Fig. 6.8) comprises a refrigerant compressor, condenser, expansion device and the vaporising section of the process cooler. In designing refrigeration circuits, compressor capability, condenser coolant temperature, and thermodynamic properties of the refrigerant must be considered, as well as energy consumption. Refrigerants should also have satisfactory environmental and safety properties, because of the possibility of refrigerant leakage. A large number of refrigerants are available with diverse thermodynamic properties (see the tables of properties for selected refrigerants in Perry and Green (1984)). A useful website for refrigerants may be found at BOC www.boc-gases.com/.

Halogenated hydrocarbons, particularly chlorofluorocarbons (CFCs), were used as refrigerants up to the early 1990s, but ozone layer depletion (especially) and global warming concerns have restricted their use, leading to development of more environmentally friendly refrigerants such as hydrochlorofluorocarbons (HCFCs) and hydrofluorocarbons (HFCs). Traditional refrigerants include $NH_3$, $C_3H_8$, $CO_2$, and (for very low temperatures) $N_2$; some of these refrigerants are toxic or flammable, and their inventories must be minimised.

Technical and operational constraints must be considered in designing the refrigeration circuit. For example, the evaporating temperature of the refrigerant should correspond to a vapour pressure above atmospheric pressure, to ensure the suction pressure of the refrigerant compressor is above atmospheric pressure. (If the suction pressure is below atmospheric pressure, there is possibility of air ingress from atmosphere into the refrigerant circuit.) Note also that the condensing temperature of the refrigerant must be greater than the temperature of the coolant used (generally cooling water) to achieve a temperature driving force for heat transfer in the refrigerant condenser.

Potential for ice formation at the tube wall (process side) of the process cooler/refrigerant vaporiser should be evaluated. If ice forms, two exchangers will be needed to allow for periodic de-icing of one of the exchangers, leading to complexity in plant operation.

Insulation of equipment and piping is needed for minimising energy loss from cold streams.

## 6.16  Electricity Demand and Supply

The total demand for electricity supply incorporates domestic, commercial and industrial needs. Within this demand, there is an underlying constant load or 'base load' demand for electricity generating capacity. This base load capacity tends to be supplied by power generation plants that operate economically at a relatively constant output at a high level of capacity utilisation. Peak demands beyond base load are normally supplied with plants of lower efficiency and lower capital cost but higher operating cost, which operate for limited periods to meet this peak demand.

Most of Australia's electricity is generated at 'base load' power stations, the majority of which use black coal or (in the case of Victoria) brown coal. Natural gas is also used. Electricity can be generated from alternators driven by steam turbines or gas turbines. Some typical fuel and steam cycles used are shown in Figs. 6.9 and 6.10. A number of generation systems are possible, with generation efficiencies depending on the fuel and generation system adopted. Table 6.7 lists typical electrical generation efficiencies for common

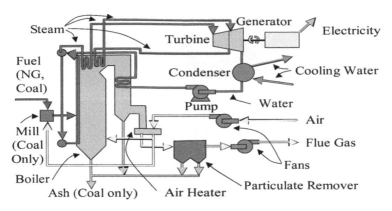

Generic Steam Turbine

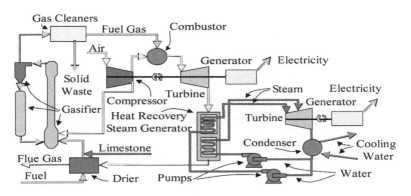

Integrated Gasification Combined Cycle

**Figure 6.9** Generic steam turbine and integrated gasification combined cycle.

fuels and the systems depicted in Figs. 6.9 and 6.10. An efficiency range is given for each system, reflecting variations in design and operation. Electrical generation efficiency may be defined as

$$\frac{\text{electrical energy exported from power station}}{\text{fuel energy consumed}}$$

The integrated gasification combined cylcle (IGCC) system converts black coal into a fuel gas containing CO, $H_2$, $CH_4$ which is combusted in a gas turbine. Exhaust gas from the gas turbine passes through a boiler to raise steam which is expanded through a steam turbine. Electricity is generated through the gas and steam turbines, similar to a natural gas combined cycle gas turbine system. The technology is under development.

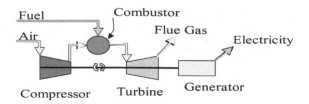

Open Cycle Gas Turbine

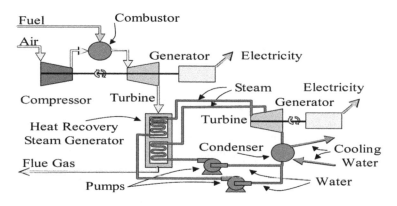

Combined Cycle Gas Turbine

**Figure 6.10**   Open and combined cycle gas turbine power cycles.

Table 6.7   Typical efficiencies for electricity generation systems

| System | Efficiency (%) |
|---|---|
| Steam turbine — black coal | 30–37 |
| Steam turbine — brown coal | 23–32 |
| Open cycle gas turbine | 17–37 |
| Combined cycle gas turbine | 44–57 |

Environmental impacts at power stations include

- emissions resulting from fuel combustion, mainly $CO_2$, $NO_x$, $SO_2$, ash
- depletion of fuel resources
- depletion of water resources due to evaporative loss in cooling towers
- impacts resulting from aqueous emissions or their treatment

- thermal energy released where steam is condensed, usually via cooling water.

Details of typical fuel consumptions and waste emissions for common generation and fuel systems are provided in Table 11.1 of Chapter 11.

Cogeneration is the combined generation of electricity and the use of exhaust steam from the turbine for process heating or other heating (e.g., domestic). Where cogeneration is adopted, efficiencies are much higher but this requires a steady demand for steam, typically from a continuous process industry. Environmental benefits of cogeneration result from

- avoiding the fuel consumption and emissions incurred in alternative dedicated steam generation for process heating
- reduced emission of thermal energy to the environment
- reduced consumption of cooling water, including reduced consumption of make-up water for a recirculated system incorporating cooling towers.

Other possible means of electricity generation include

- hydroelectric
- nuclear fission
- solar
- wind
- geothermal.

These means of generation, as well as others under development, are reviewed in Hardin (2008). Hydroelectric and nuclear fission play an important contributory role in many countries. Solar, wind and geothermal approaches offer much potential in terms of greatly reduced demand on resources and emissions. These sources are currently under development, but as yet provide only a limited share of generating capacity in many countries.

## 6.17 Distribution and Use of Electricity

It is common to distribute electricity from a power station at high voltages, for example, 11 kV, and transform to lower voltages (3.3 kV, 415 V, 240 V) at a process plant. Power supplied at 415 V is available as 3-phase power at a frequency of 50 Hz in Australia. There are efficiency losses in electric power distribution from the power station to the user plant which increase with increased transmission distances. The frequency of electricity supply differs in some other countries, for example, the United States, where it is 60 Hz.

Electricity is used in the process industries for

- electrochemical processes (e.g., aluminium smelting, chlorine/caustic soda plants) where transformer rectifiers convert AC power to DC power at required voltages
- driving machinery within process plants, especially compressors, pumps, agitators, solids conveyors, centrifuges, size reduction equipment
- instrumentation, computer systems, lighting.

## 6.18  Compressed Air

Various factory uses of compressed air demand different quantities, qualities, pressures and supply reliabilities, for example,

- instrument air should be free of oil and condensed moisture
- safety air must be suitable for breathing
- process air is used for purging toxic gases, emergency agitation
- maintenance air is used for air-driven tools.

Options exist in terms of separate or integrated compression and supply systems.

## 6.19  Inert Gas

Inert gas, typically nitrogen, is required for purging flammable or explosive gases or rendering gas spaces inert from flammability or explosion ('inerting'). Important properties of nitrogen for this purpose are

- pressure
- moisture content
- oxygen content ($<0.1\%$)
- reliability of supply.

Supply options for nitrogen include

- local generation
- bulk storage
- pipeline.

## 6.20 Vacuum

Vacuum is often used in filtration, and for heat sensitive materials in evaporation and distillation. Vacuum can be achieved by pumps or steam jet ejectors. Initially the system is evacuated to reach the desired vacuum; the vacuum is then maintained by removing gases resulting from air ingress or evaporation of the process fluid.

The size of a vacuum pump depends on

- absolute pressure (extent of vacuum) required
- volume of system evacuated
- nature, quality of vapours/gases to be handled
- evacuation (or 'pump down') time.

Steam jet ejectors can achieve progressively higher vacuums by using multiple ejectors in series. Advantages of steam jet ejectors include

- absence of valves, rotors, pistons, moving parts
- ability to be constructed in a wide range of materials.

Condensers are used in conjunction with vacuum systems to condense evaporated liquids. The consequences of handling condensed liquids from vacuum systems must be considered. Aspects include condensate recovery from steam jet ejectors, purification of contaminated condensate, and recovery of process materials from condensed liquids. In vacuum distillation of heavy oil in petroleum refineries, vacuum pumps are often preferred to steam jet ejectors in order to reduce the contamination of condensed water by oil.

## 6.21 Concluding Remarks

Utilities play an essential and integral part in processes, principally through supply and use of energy. Utility supply must be highly reliable and meet quality demands which in many cases are stringent.

The generation of utilities draws on fuel and water resources and leads to a number of emissions contributing to a range of environmental impacts. Aggregated over a product's life cycle, utilities make a major contribution to the environmental impact of the product. Accounting for impacts derived from utility use is an important element of environmental assessment; this is explored in Chapters 10 and 11 in the context of life cycle assessment methodology. Utilities also impose capital and operating cost burdens which must be borne directly or indirectly by the processes consuming the utilities; these aspects are discussed further in Chapter 13. In Chapter 7, we explore some common ways in which utilities are used, and how utility consumption in process plants can be reduced.

## References

Cohen Hubal, E. A. and Overcash, M. R. (1993) Waste reduction analysis applied to air pollution control technologies, Air Waste, 43, 1449–1453.

Felder, R. M. and Rousseau, R. W. (2005) *Elementary Principles of Chemical Processes*, 3rd edn, Wiley, Hoboken, New Jersey.

Fontana, M. G. (1986) *Corrosion Engineering*, McGraw-Hill, New York.

Perry, R. H. and Green, D. (1984) *Perry's Chemical Engineering Handbook*, 6th edn., McGraw-Hill, New York.

Hardin, M. (2008) *The Start of the Road: Energy and Chemical Engineering*, A Report for the Institution of Chemical Engineers, IChemE, Rugby, UK.

Sinnott R. and Towler G. (2009) *Chemical Engineering Design*, Elsevier, Oxford, UK.

Smith, R. (2005) *Chemical Process Design and Integration*, Wiley, Chichester, England.

# Chapter 7

# Energy Conservation

## 7.1 Introduction

Apart from improving the design and operation of utility systems, a key strategy for reducing environmental impacts derived from utility use is to reduce utility consumption on process plants.

Gas compressors and liquid pumps have a key role in most chemical processes by conveying fluids. Compressors and pumps are major consumers of energy, commonly supplied as electricity, but also as steam. Energy consumed is a function of the flow rates and pressures of process streams, but is also influenced by pressure losses within equipment and piping. We first consider selection of compressors and pumps for process duties and their energy requirements. We then explore contributions to pressure loss in equipment and piping.

Electrochemical processes such as chlorine manufacture and aluminium smelting depend on DC electricity supply. In electrochemical processes, improved reactor designs, often involving fundamental changes in technology and improved process control are key to reducing energy consumption.

Consumption of hot and cold utilities in process plants can be reduced by efficient heat recovery from process streams, and by effective insulation on equipment and piping.

*Sustainable Process Engineering: Concepts, Strategies, Evaluation, and Implementation*
David Brennan
Copyright © 2013 Pan Stanford Publishing Pte. Ltd.
ISBN 978-981-4316-78-1 (Hardcover), 978-981-4364-22-5 (eBook)
www.panstanford.com

## 7.2 Energy Consumption in Compression of Gases

In many chemical processes (e.g., ethylene, chlorine, and sulphuric acid manufacture) a gas compressor handles the main process stream. Compressor performance in these cases is vitally important in terms of reliability, cost (capital and operating), and energy consumption. There is a terminology progression from *fans* to *blowers* to *compressors*. The terms apply to the different pressure rise (delivery pressure minus suction pressure); *fans* have a small pressure rise, *blowers* somewhat larger, *compressors* larger again. We now consider key aspects of compressor specification and selection.

### 7.2.1 *Process Specification for Gas Compressors*

A range of gas properties for both suction and delivery conditions influence machine selection for the compression duty as well as energy requirements.

Suction conditions normally specified are

- temperature (K)
- pressure (kPa abs)
- volumetric flow rate (m$^3$/h)
- composition.

Delivery condition normally specified is

- pressure.

Other important gas properties include

- the ratio of specific heats $\gamma = C_p/C_v$, which influences the energy consumed and gas temperature increase during compression
- Interaction with lubricant, for example, the cases of $O_2$, $Cl_2$
- density
- molecular size, which influences gas leakage within compressor clearances and at compressor shaft seals
- corrosivity affecting materials selection
- viscosity
- contaminants, including suspended solid or liquid particles
- toxicity, which has implications for sealing against leakage
- flammability, which has implications for sealing against air ingress.

### 7.2.2 *Machine Selection*

Key performance parameters affecting machine selection include

- **volumetric capacity**, traditionally expressed in $m^3/h$ referred to suction conditions

- **compression ratio** $= \dfrac{\textbf{discharge pressure}}{\textbf{suction pressure}}$

where pressures are expressed in absolute units.

There is a distinction here from pumps, where performance is characterised by volumetric capacity and **developed head** = **delivery pressure** − **suction pressure.**

Another key consideration is whether the machine is lubricated or non-lubricated within the gas compression chamber. Some gases (e.g., chlorine) react with lubricating oil to produce a non-lubricating solid which fouls the compression chamber surfaces. Oxygen supports combustion and hence oxygen should be isolated from lubricating oil. Contamination from lubricants may not be acceptable in certain downstream equipment (e.g., in some catalyst beds).

There is a range of compressor types and machine compression mechanisms, but compressor performance can be classified as one of two basic forms, that is,

- positive displacement or
- aerodynamic (sometimes referred to as 'dynamic').

Performance curves of pressure versus flow rate are fundamentally different for each compressor type. Figure 7.1 shows typical performance curves for centrifugal and axial flow (aerodynamic) and reciprocating (positive displacement) machines.

A requirement of process gas compressors is to handle varying gas flow rates to match different production rates and plant operating conditions. For a *positive displacement machine* operating at a given speed, a fixed flow rate is delivered independent of compression ratio. To vary gas flow rate through the machine, the machine speed must be changed; in some machines, the displacement volume can be modified.

For an *aerodynamic machine* operating at a fixed speed, the gas flow rate is dependent on the compression ratio, and can be manipulated by adjusting valves on the compressor suction or delivery. Gas flow rate can also be manipulated by adjusting stationary vanes on the compressor inlet; *this approach has the advantage of lowering energy consumption.* Note here an important distinction from centrifugal pumps, where suction valves are never throttled because of cavitation concerns. Gas flow rate can also be modified by changing machine speed.

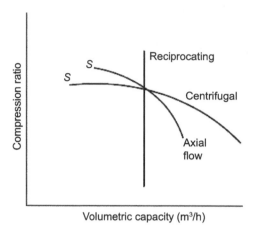

**Figure 7.1**   Comparison of compressor performance curves ($S$ denotes surge point).

There is a lower limit to gas flow rate through an aerodynamic compressor, called *surge point*, where gas flow and pressure development are unstable. Operation at gas flow rates below the surge point is not practicable.

For an indication of the range of compressor machine types, and their performance range, see Chapter 10 in Sinnott and Towler (2009) or similar texts. Note the different speeds for various machine types, for example, slow for reciprocating (~300 rpm), fast for screw compressors (positive displacement) or centrifugal compressors (aerodynamic) (~10,000 rpm). For a given duty, low speed reciprocating machines are larger and require larger foundations and more space. *Also, note that there is a range of achievable discharge pressures and flow rates for different machines, limiting machine choice for a given duty.*

## 7.2.3  *Thermodynamics of Gas Compression*

Compression can approach

- adiabatic (more common) or
- isothermal

depending on machine type and extent of cooling during compression. Isothermal compression is approached in a liquid ring compressor; adiabatic compression is approached in centrifugal and most reciprocating machines. Departures from true adiabatic conditions occur, for example, due to cylinder cooling in reciprocating machines, and gas slippage and recycling in screw compressors.

Design and thermodynamics texts provide equations for different types of compression in terms of

- theoretical compressor work
- discharge temperature.

for single-stage and multistage compression.

For single-stage *adiabatic compression*, theoretical work $W$ is given by

$$W = \left(\frac{\gamma}{\gamma - 1}\right) P_1 V_1 \left[\left(\frac{P_2}{P_1}\right)^{\frac{\gamma - 1}{\gamma}} - 1\right] \tag{7.1}$$

while temperature increase is given by

$$\frac{T_2}{T_1} = \left(\frac{P_2}{P_1}\right)^{\frac{\gamma - 1}{\gamma}} \tag{7.2}$$

where $T_2$, $T_1$ are absolute temperatures; $P_2$, $P_1$ are absolute pressures; and $V$ is volumetric flow rate and $\gamma$ is the ratio of specific heats $C_p/C_v$. Subscripts 1 and 2 denote suction and delivery conditions, respectively.

The power consumed by a compression machine

$$= \frac{\text{theoretical work}}{\text{compressor efficiency}}$$

Compressor efficiency is function of

- machine type
- compression duty
- properties of the gas being compressed.

In the absence of detailed information, a first assumption of 75% is reasonable for compressor efficiency. (Sinnott and Towler, 2009 show typical efficiencies for centrifugal, axial flow and reciprocating compressors.)

**Case example**
A gas compressor is required to compress 250 m$^3$/h methane supplied at 25°C and 1 bar abs to a pressure of 3 bar abs. If compression is adiabatic and performed in a single stage, calculate the theoretical power requirement, and the discharge temperature. Estimate the size (kW) of an electric motor for the compressor drive. Value of $\gamma$ for methane = 1.31.

Theoretical power requirements for compression are estimated as

$$W = \frac{100 \times 250 \times 1.31[3^{0.31/1.31} - 1]}{3600 \times 0.31}$$

$$= 8.7 \text{ kW}$$

Discharge temperature is estimated as

$$T_2 = 298 \times 3^{0.31/1.31}$$

$$= 386.5 \text{ K} = 113°\text{C}$$

Allowances must be made for both compressor and motor efficiencies. If we allow 75% for compressor efficiency and 95% for motor efficiency, absorbed power becomes $8.7/(0.75 \times 0.95) = 12.2$ kW. For motor selection, the next standard size of electric motor available would be selected.

### 7.2.4 *Limits to Compression Ratio per Stage of Compression*

Limits to compression ratio $(P_2/P_1)$ are imposed by machine types, but can also be imposed by maximum allowable gas temperatures. For example, chlorine will react with steel at around $130°$C, thus limiting discharge temperatures.

### 7.2.5 *Intercooling of Gas during Compression*

Intercooling of gas is needed in multistage compression to minimise overall gas temperature rise and overall power consumption. Power consumption is proportional to inlet gas volume, which in turn is proportional to absolute temperature of the gas. The number of stages and extent of intercooling is constrained by capital cost considerations.

### 7.2.6 *Reliability*

Some machines (e.g., centrifugal, screw) can run continuously for long periods (e.g., 2–3 years) without shut down. Other machines (e.g., reciprocating) must be taken out of service for maintenance after shorter periods (e.g., 9–12 months). To ensure process reliability, the question arises regarding duplication of machines and their capacities. Options might include

- one machine of 100% capacity
- two machines each of 100% capacity
- two machines each of 50% capacity.

Such options must be weighed up in the light of required performance, reliability, and costs.

## 7.2.7  Drives for Compressors

Possible drives for gas compressors include

- electric motors
- steam turbines (for rotating compressor shafts)
- (less commonly) gas turbines, other combustion engines.

Drives are available in standard sizes. The nominal power rating for the drive must exceed the maximum absorbed power of the compressor in operation, including abnormal as well as design duties. Drives and drive assemblies have limited efficiencies. Drive efficiencies tend to increase with size of drive. Steam turbine efficiencies are lower than electric motor efficiencies, and depend on capacity utilisation; drive efficiencies tend to be higher for larger machines running at higher capacity utilisation.

## 7.2.8  Energy Conservation in Gas Compression

For a given gas compression duty within a process, a number of areas can be targeted for reducing energy consumption. These include

- modifying the process conditions, where possible, to reduce power consumption
- reducing pressure losses in equipment and piping
- improving the efficiency of the compressor
- improving the efficiency of the compressor drive
- choosing an energy efficient method of capacity control, including the selection of variable speed drive.

## 7.3  Energy Consumption in Pumping of Liquids

### 7.3.1  Process Specification for Pumps

As for compression, we first consider the properties of liquids pumped. Important properties and their effects are listed below.

| Liquid properties | Effects |
|---|---|
| Newtonian or non-Newtonian | Liquid behaviour under shear stress |
| Density | Power consumed |
| Viscosity | Head, efficiency, power consumed |
| Composition and contaminants (gases, dissolved solids, suspended solids) | Corrosivity |
| Corrosive properties | Materials selection, speed selection |
| Temperature | Materials selection, vapour pressure |
| Vapour pressure | Net positive suction head (NPSH) |

Some key performance parameters for pumps include

**Volumetric capacity** — range of volumetric flow rates ($m^3/h$) required

**Suction conditions**

- pressure (at pump suction flange)
- temperature
- vapour pressure and NPSH consideration

**Delivery pressure** — pressure of liquid delivered at pump delivery flange

**Developed head = Delivery pressure − suction pressure**

The developed head of a pump, usually expressed in metres of fluid, is shown on the pump performance curve, plotted against volumetric flow rate.

**Net positive suction head (NPSH) = Suction pressure − vapour pressure**

Sufficient positive pressure must be provided to the pump suction to avoid vaporisation and consequent cavitation damage within the pump. Cavitation results from collapse of vapour bubbles on metal surfaces within the pump. Liquid at the pump suction flange may vaporise before reaching the eye of the impeller due to internal pressure drop within the pump. This pressure drop is difficult to calculate, because of geometric complexity and change of fluid direction within the pump. Hence it is measured and specified by the pump manufacturer as a net positive suction head required (NPSHR) at the pump suction flange. NPSHR increases with increasing volumetric flow rate for a pump of fixed speed, and also with increasing pump speed. The designer must ensure that the available NPSH (pressure at pump suction − vapour pressure) exceeds the NPSHR specified by the pump manufacturer. Vessels on the suction of a pump must often be elevated to provide adequate suction head. Liquids have high vapour pressures when they are close to their boiling point, for example, distillation column bottoms streams, boiler feed water, liquefied gases from storage vessels.

## 7.3.2 *Power Requirement*

The power $P$ required at the pump shaft (usually expressed in kW) can be calculated as

$$P = \frac{V \times \Delta h \times \rho \times g}{\eta'}$$

where    $V$ = volumetric flow rate

$\Delta h$ = developed head

$\rho$ = liquid density

$g$ = acceleration due to gravity

$\eta'$ = pump efficiency (ratio of theoretical power to actual
power at pump shaft)

Pump efficiency depends on pump type and its operating duty point, but is typically in the range of 60%–80%. To determine the required power of the driver, allowance should be made for the efficiencies of the drive, and the coupling system connecting pump and drive shafts.

**Case example**

A pump is required to deliver 154 kg/s of water from a suction head of 10.7 m fluid to a delivery head of 32.0 m fluid. Heads are inclusive of system pressure, static head and friction losses. Developed head $= 32.0 - 10.7 = 21.3$ m

Assuming 70% efficiency for combined pump and motor efficiency, power required for pumping $= \dfrac{154 \times 21.3 \times 9.8}{0.7 \times 1000} = 46\,\text{kW}$

## 7.3.3 *Pump Machine Types*

Pumps may be classified as

- centrifugal or
- positive displacement machines.

Common examples of positive displacement pumps include

- rotary gear pump (often used for liquid fuels in combustion systems)
- piston pump (often used for dosing chemicals to steam boilers).

As with compressors, there is a distinction in performance curves for

- positive displacement, where flow rate at a given pump speed is independent of head
- centrifugal, where flow rate at a given pump speed is dependent on developed head.

### 7.3.4 *Centrifugal Pump Selection and Performance*

For a given speed, operating ranges of flow rate and head for various sized pumps are shown on a pump selection chart provided by a pump supplier. Performance curves for a specific pump at a given speed are usually provided as a family of curves of developed head (metres fluid) versus volumetric flow rate (m$^3$/h) for different diameter impellers (Fig. 7.2). A curve for NPSHR (m fluid) versus volumetric flow rate (m$^3$/h) is usually also provided.

The curves are independent of fluid density, but not of fluid viscosity; higher viscosity liquids reduce the developed head and efficiency at a given

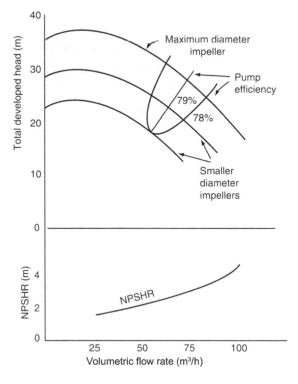

**Figure 7.2** Example of performance curves for a centrifugal pump at a given speed.

flow rate. Power consumption curves are sometimes shown, but these are usually for water and must be adjusted for the density of the liquid pumped.

Where the pump is directly coupled to a *fixed speed* electric motor (a common arrangement) pump speed is the same as the motor speed. The speed of a synchronous electric motor is related to the frequency ($f$) of power supply and the number (always even) of poles ($p$) in the motor by $N$ (rpm) $= 120 \times f/p$.

Thus for $f = 50$ Hz,

| $p$ | $N$ (rpm) |
|---|---|
| 2 | 3000 |
| 4 | 1500 |
| 8 | 750 |

Actual speeds are usually slightly less than predicted due to slippage effects in the motor. Note that operating pump speeds may differ from one country to another reflecting the frequency of power supply. For example, $f = 50$ Hz in Australia, but 60 Hz in United States of America.

Variable speed motors offer potential for reduced energy consumption (Ruddell, 2010). Such motors are usually higher in capital costs, but lower in operating costs, and offer environmental benefits derived from reduced energy usage.

Higher pump speeds generate higher flow rates and heads, usually have higher efficiencies, and are smaller in size than pumps providing the same performance at lower speeds. This offers potential cost saving, though pump speed may be limited by corrosion and wear effects on pump internals. For example, slurry pumps typically run at lower speeds (1000 rpm or less) than clean water pumps (1450 or 2900 rpm).

Dimensional analysis gives a guide to effect of speed on performance, that is,

$$V \quad \propto \quad ND^3$$
$$\Delta h \quad \propto \quad N^2 D^2$$
$$P \quad \propto \quad \rho N^3 D^5$$

where $N$ = pump impeller speed, $D$ = impeller diameter. Multistage pumps can achieve higher developed heads, for example, those required for boiler feed water duty. Pumps, like compressors, may be driven by electric motors or steam turbines. As with compressors, pump reliability is important in process plants. It is common practice to duplicate pumps to achieve the required reliability; this can usually be justified because of lower capital costs than for compressors.

### 7.3.5  *Energy Conservation in Pumping of Liquids*

Energy conservation opportunities include

- selecting a pump with higher efficiency
- modifying the process duty, for example, by reducing the required head through equipment design, pipe sizing, and plant layout decisions
- using variable speed drives.

## 7.4  Pressure Losses in Piping

Energy consumption through pumping and compression of fluids in process plants may be reduced by reducing pressure losses in piping and equipment. Pressure losses in piping for a given fluid system are influenced by the diameter, length, design and layout of piping, and the selection and design of valves and other pipe fittings.

### 7.4.1  *Sizing of Pipes*

In selecting pipe sizes for process and utility streams, several considerations are important. Piping makes a large contribution, typically 20%, to process plant capital costs. Piping cost increases with increased piping lengths, and hence the complexity and layout of the plant, but also with increased pipe diameter. There is usually a trade-off between capital cost (favoured by smaller diameter) and energy consumed in pressure loss (favoured by larger diameter). The trade-off between capital and operating costs is the basis of so called 'optimum diameter' equations. Optimum diameter equations are based on several assumptions (economic and technical), which may not apply to all design cases.

Rules of thumb are widely adopted in initial sizing of pipes. For example, for a low viscosity liquid, a velocity of 2 m/s might be adopted while for a gas at low pressure, 15 m/s might be adopted. For gases at higher pressures, higher velocities can be tolerated because of higher allowable pressure drops. Sinnott and Towler (2009) provide useful guidelines for sizing pipes based on superficial velocities, with corresponding pressure drops per length of pipe. For viscous liquids, a range of superficial velocities based on pipe diameter and liquid viscosity is recommended.

There are *other factors* influencing pipe sizing which include accounting for

(i) erosion or corrosion, for example,

    (a) when handling 98% sulphuric acid at ambient temperature in mild steel or cast iron piping, the velocity is often limited to 1 m/s to minimise corrosion

    (b) tube side velocity for cooling water in shell and tube heat exchangers is often limited to 2–2.5 m/s to minimise erosion of metal tubes

(ii) effects of fouling

    deposition of scale, corrosion products, or suspended solids decreases the effective pipe diameter. For a given mass flow rate, pressure drop is inversely proportional to $d^5$, where d is the internal pipe diameter, so the effect of fouling can be significant. This effect is also relevant to heat exchanger tubes.

Pipes are constructed in *standard sizes* — see Table 6.6 in Perry and Green (1984) for steel and stainless steel pipe. Branch sizes on process equipment in most cases will be similar to connecting pipe sizes.

## 7.5 Pressure Loss through Equipment

### 7.5.1 *Heat Exchangers*

Heat exchangers are widely used for heating and cooling process streams in process plants. Duties may involve using hot or cold utilities, but may alternatively involve exchanging heat between hotter and colder process streams. Heat exchangers are also used to recover energy from process streams to generate hot or cold utilities. The rate of heat transfer is given by

$$Q = U A \Delta T$$

where    $Q$ = rate of heat transfer

            $U$ = overall heat transfer coefficient

            $A$ = surface area for heat transfer

           $\Delta T$ = overall temperature driving force for heat transfer

The rate of heat transfer must also satisfy the energy balance for the hot and cold streams entering and leaving the heat exchanger, which may be undergoing sensible and/or latent heat changes.

$U$, the overall heat transfer coefficient is a function of individual heat transfer coefficients for each process stream and any fouling occurring on heat exchanger surfaces. Values of individual heat transfer coefficients reflect the mode of heat transfer occurring (e.g., condensation, forced convection) and the properties of the fluids comprising the process streams. Higher values of $\Delta T$ will reduce the surface area requirements and hence capital cost of an exchanger, but selection of utility and process stream temperatures will be constrained. Careful selection of terminal temperatures for process streams and utilities is an important aspect of heat exchanger design and operation. This is a consideration in tackling all three of the flow-sheeting problems in Part D of this book.

Heat exchangers contribute to energy consumption arising from hot and cold utility use, but also from pressure loss due to fluid flow through the exchanger which contributes to pumping or compression energy. Pressure drops increase with higher velocities and longer flow paths of fluids. Certain types of heat transfer (e.g., forced convection) are aided by turbulence and higher velocities, which in turn lead to higher fluid pressure drops through the exchanger. In these cases there is a trade-off between capital costs and pressure drop in exchanger selection and design. A typical rule of thumb for pressure drop in an exchanger is 30 kN/m$^2$ for low viscosity liquids; target pressure drops for gases reflect the supply pressure of the gas. Design considerations for heat exchangers are covered in Sinnott and Towler (2009) who also provide guidelines for typical heat transfer coefficients and pressure drops.

### 7.5.2 *Vapour–Liquid Contacting Columns*

More difficult separations in distillation and gas absorption require greater packing heights, or more trays, contributing to higher column pressure drops.

The height and diameter of a *distillation column* for a given separation can be modified by changing the *reflux ratio*, as discussed in Section 5.3. Higher reflux ratios reduce the number of trays required but at the expense of increased utilities consumption.

For vapour–liquid contacting, a choice can be made between *tray* and *packed* columns. The primary purpose of *trays* in a tray column is to ensure good vapour–liquid contact for mass transfer, but different tray types and designs incur different pressure drops. For example, a bubble cap tray typically has a higher pressure drop than a sieve tray for a given vapour–liquid flow regime because of its relative vapour pathway, even though it may offer compensating advantages. For a given tray type, choice of parameters such as the number of vapour openings per unit of tray area (affecting vapour velocity), and liquid hold-up on the tray, will affect vapour pressure drop through the tray. Allowable tray pressure drops are strongly influenced by system pressure; for example, vacuum distillation will impose more stringent pressure drop requirements. As a guide, the dynamic slot seal (mean head of liquid above top of slots in the cap) for bubble cap trays might range from 20–25 mm for vacuum conditions to 50–75 mm for moderate pressure conditions (Bolles, 1963).

Selection of packing size and type for a packed column influences the vapour pressure drop through the column. Pressure drops for conventional packing types typically vary between 0.2 and 1.2 kPa/m of packing height, while structured packing might offer as little as 0.1 kPa per metre of packing. In addition to the pressure drop through the packing, column internals such as packing support plates, liquid distributors, mist eliminators, and hold-down plates all contribute to overall column pressure drop. The designs of liquid distributor and packing support plate can be modified to minimise vapour pressure drop.

## 7.6 Agitation and Mixing

Agitation is provided in vessels to promote fluid mixing. Mixing objectives can be diverse, and in some cases multiple. Some common objectives include:

- maintaining uniform temperature or concentration in a reactor
- blending of two miscible liquids
- suspension of solids in a liquid
- dispersion of gas in a liquid
- dispersion of one immiscible liquid in another liquid
- promotion of heat transfer from a jacket or coil.

Mixing in vessels can be achieved by rotational impeller(s) on a shaft, or by jets. Impellers draw fluid into the impeller zone and discharge it in a particular direction. The resulting fluid motion depends on the magnitude and type of disturbance to the fluid and the physical properties of the fluid, especially viscosity. Selection of impeller type, diameter, and speed for a mixing application must consider the purpose of mixing, the phases present, and fluid viscosity. Jets create mixing by entraining fluid and achieving fluid circulation patterns; jet mixing is often used in large storage tanks.

Impellers vary in shape and size, but can be broadly classified as

- propellers — high speed, promoting predominantly axial flow
- turbines — moderate to high speeds, promoting mainly radial flow
- paddles — low speed promoting predominantly tangential flow.

Typically turbines and propellers have small ratios of impeller to vessel diameter ($D/D_T$), operate at high speeds, and in low viscosity liquids. Paddles have high $D/D_T$ ratios and are used in high viscosity liquids or where gentle agitation is required, for example, when avoiding physical damage to crystals.

*Agitated vessels* vary in geometry and design. Vertical, cylindrical vessels are most widely used. Impeller mounting is usually vertical at the vessel axis with top shaft entry, but bottom shaft entry at vertical axis, and side entry near the base of large tanks are also practised. Vessels can have various internals affecting mixing, for example,

- baffles
- coils for promoting heat transfer
- pipes – for example, for liquid feed, probes – for instruments
- baskets (false bottoms) for dissolving solids.

*Baffles* are used to eliminate swirl and promote radial and axial motion in the vessel. Centrally mounted impellers in the absence of baffles promote tangential motion, with vortices formed in low viscosity liquids. Vortex depth can be predicted from velocity profiles and pressure balances (Brennan, 1976).

*Agitator performance* has traditionally been evaluated using dimensional analysis in conjunction with scale-up, reflecting the diversity of mixing objectives, vessel and impeller geometries, and the complexity of turbulence. Computational fluid dynamics is increasingly applied to assist in design. Experimental studies at pilot scale have focused on

- fluid circulation produced by the impeller
- power consumed by the impeller
- performance criteria related to the mixing objective, for example, mixing time or heat transfer rate.

*Power input* for agitated vessels can vary widely, ranging between 0.01 and 2.0 kW/m$^3$ of fluid volume depending on the mixing objective and the severity of agitation (Sinnott and Towler, 2009). Correlations for power consumption at the mixer shaft have been developed for vessels with standard geometries (typically, $D/D_T = 1/3$, and four vertical wall baffles each of width 0.1 $D_T$). Correlations have used dimensionless group plots of Power number ($P/\rho N^3 D^5$) versus Reynolds number ($ND^2 \rho/\mu$) where $\rho$ and $\mu$ are fluid density and viscosity, respectively. $ND$ is taken as a characteristic velocity of the impeller with $N$ the rotational speed (in number of revolutions per unit time) and $D$ the impeller diameter. Additional allowance must be made for drive and coupling efficiencies.

For impeller Reynolds numbers (Re) between 1 and 10, Power number $N_p$ is inversely proportional to Re. This region is referred to as the **laminar region.** For Re $> 10, 000$, $N_p$ for a given impeller is independent of Reynolds number. This region is referred to as the *turbulent region.* For Reynolds numbers between 10 and 10,000 there is a *transition region.* For low viscosity liquids, impeller speed should be chosen in most cases to ensure mixing in the turbulent region.

Correlations have also been developed for mixing performance, for example, *mixing time* $\theta$ taken in a single-phase batch system to achieve a defined degree of homogeneity; for baffled vessels in the turbulent region, N$\theta$ = constant. Where *heat transfer* occurs, heat transfer correlations resemble those for heat transfer in pipes. The overall heat transfer coefficient is estimated from the impeller heat transfer coefficient, and external coefficient for the jacket fluid.

Useful references on mixing in agitated vessels include Sinnott and Towler (2009), Perry and Green (1984), and Harnby, Edwards, and Nienow (1985). An important consideration in all mixing operations is to use the power input efficiently to achieve the desired mixing performance. A variable speed drive, usually incorporating a gearbox, offers flexibility in selecting the optimal speed for mixing, and opportunity for energy conservation.

Static mixers used for mixing gaseous reactants prior to the reactor were identified in Chapter 4 as having an important role in waste minimisation. Static mixers are also used for a wide range of liquid mixing applications and are an alternative to using agitation in vessels.

## 7.7 Heat Recovery

Consumption of hot and cold utilities on process plants can be minimised by making best use of heat recovery opportunities from process streams. These opportunities include heat recovery from hot streams to generate steam and from very cold streams to generate general purpose coolants.

Opportunities also include exchange of heat between selected hot and cold process streams to reduce overall demand in the plant for hot and cold utilities. To maximise these opportunities, careful attention to temperature profiles and temperature driving forces for plant heat exchangers is necessary. A simple example is the recovery of process heat from a high temperature gas stream by generating steam in an economiser, boiler, and superheater. Heat from the gas stream is exchanged counter currently with the water/steam stream, with the hottest gas cooled in the superheater and the final cooling in the economiser.

A further example is evident in the simplified flow diagram of sulphuric acid manufacture (see Fig. 4.5). Gas leaving the third pass of the converter at around 450°C requires cooling to a lower temperature suitable for $SO_3$ absorption. This gas stream is first cooled by exchanging heat with gas leaving the intermediate absorber, and then further cooled by raising steam in a waste heat boiler prior to entering the intermediate absorber. The gas leaving the intermediate absorber undergoes two stages of heating before returning as feed to the fourth stage of the converter: firstly by the gas feed to the absorber, and secondly by the gas stream from the second stage of the converter.

Much attention has been given in recent decades to the use of heat integration techniques. These are discussed in design texts, for example, at an introductory level in Sinnott and Towler (2009) and at a more advanced level in Smith (2005). Process integration and associated pinch technology are particularly important where large numbers of heat exchangers are required, such as in petroleum refineries and chemical plant complexes. The design approach can be used to minimise utility consumption and capital cost of the heat exchanger network.

## 7.8  Energy Recovery from High Pressure Streams

Energy can be recovered from high pressure gas or vapour streams using an expansion turbine (or 'turboexpander'). The energy thus recovered can be used to drive gas compressors. The expansion of most gases through a turboexpander produces a cooling effect in the expanded gas (Joule–Thomson effect), although there are exceptions (e.g., in the case of hydrogen, where a heating effect results). Turbo expanders are widely used in the refrigeration and liquefaction of gases where gas cooling is required, with the added benefit of energy recovery for gas compression. Common applications are in the processing of natural gas for the extraction of ethane and liquefied petroleum gas, and in the liquefaction of natural gas to produce liquefied natural gas.

## 7.9  Insulation

Where temperatures of process or utility streams differ from ambient conditions, temperature losses from equipment and piping can be minimised by insulation. Heat losses by conduction through the insulating material are proportional to the thermal conductivity and inversely proportional to the thickness of the insulating material. Actual losses will exceed calculated losses due to factors such as moisture ingress, provision of expansion–contraction joints, or poor practice in installation or maintenance. Choice of insulating material depends on the operating temperature; perlite is widely used at low or cryogenic temperatures, while calcium silicate might be used at around 500°C. Chapter 11 in Perry and Green (1984) discusses insulation materials and economic trade-offs between installed cost and energy savings, and provides useful references.

## 7.10 Plant Layout

There are many considerations in plant layout, but some design decisions will affect energy consumption. A more widely spaced layout will involve greater piping lengths and greater pressure drops for both process and utility piping. Longer piping lengths also provide greater opportunities for heat losses to atmosphere. Elevation of equipment will influence static head requirements for pumps. One decision in this context is the elevation of an overhead condenser and reflux drum in distillation (see Fig. 5.1). Elevation can avoid the need for a reflux pump if reflux returns to the tower by gravity; however, increased structural support will be required for both condenser and reflux drum, and the required cooling water pressure at ground level will be increased.

## 7.11 Concluding Remarks

Energy consumption in a plant is a function of many influences, including

- process and plant complexity, including number of equipment items
- equipment design features
- pipe sizing
- plant layout
- operating temperature and pressure of process equipment
- insulation practices.

An appreciation of the characteristics of pumps, compressors, agitators, and their drives, and the contribution of equipment sizing and design to energy consumption is integral with initiatives to minimise energy consumption.

Similarly, opportunities for heat recovery can greatly reduce consumption of hot and cold utilities.

Within the chemical processes reviewed in Chapter 4, gas compression is an important source of energy use in chlorine, ethylene and sulphuric acid manufacture. Heat recovery opportunities occur around reactors in ethylene, sulphuric acid and hydrogen cyanide manufacture. Agitation and cooling play an important role in reactors in PVC manufacture.

In many cases, there are trade-offs between increased investment cost and reduced energy consumption, offering reductions in both operating cost and environmental impact. An important aspect of process engineering is the perception, evaluation, and resolution of these trade-offs.

In conclusion, a checklist (Table 7.1) is provided of energy conservation opportunities discussed in Chapter 7 for the process plant. These opportunities are useful additions to those in reaction systems (Chapter 4) and selection of fuel and utility systems (Chapter 6).

Table 7.1   Seven essential energy conservation strategies for sustainable process engineering

---

**Increase process heat recovery**
between process streams to generate utilities from energy in process streams, and reduce hot and cold utility consumption
**Select energy efficient**
- compressors and pumps for process duties
- capacity control systems
- drives

**Reduce pressure drop** through equipment items and piping while cognisant of performance and cost constraints
**Achieve maximum process performance per unit of energy** for equipment such as vapour-liquid contacting columns, heat exchangers and mixing devices
**Provide effective insulation** on hot and cold equipment and piping
**Include energy conservation** as one of multiple criteria **in plant layout**
**Include environmental criteria** in assessing and resolving trade-offs between capital costs and energy consumption in design decisions

---

# References

Bolles, W. L. (1963) Tray hydraulics. Buble-cap trays, in *Design of Equilibrium Stage Processes* (ed. Smith, B. D.), Mcgraw-Hill, New York.

Brennan, D. J. (1976) Vortex geometry in unbaffled vessels, *Trans. IChemE A.*, 54(4), 208–216.

Harnby, N., Edwards, M. F., and Nienow, A. W. (1985) *Mixing in the Process Industries*, Butterworths, London, UK.

Perry, R. H. and Green, D. (eds) (1984) *Perry's Chemical Engineering Handbook*, 6th edn, McGraw-Hill, New York.

Ruddell, S. (2010) Driving to reduce emissions, *Chem. Eng.*, 824 (February), 52–53.

Sinnott R. and Towler G. (2009) *Chemical Engineering Design*, Elsevier, Oxford, UK.

Smith, R. (2005) *Chemical Process Design and Integration*, Wiley, Chichester, UK.

# Chapter 8

# Materials Recycling

## 8.1 Introduction

In this chapter, materials recycling is explored within the following contexts:

- recycling of process streams within chemical processes
- economic and environmental benefits and burdens of recycling
- closed loop and open loop recycling
- materials recycling within product life cycles
- waste treatment, landfill, and incineration, as less preferred alternatives within the waste management hierarchy.

## 8.2 Recycling of Materials in Chemical Processes

Recycle of materials is common in most processes involving chemical reactions where feedstock conversion is incomplete. Examples of such processes reviewed in Chapter 4 include

- recycling of depleted brine to brine cell feed in chlorine manufacture (Fig. 4.1)
- recycling of unreacted ethane to reactor feed in ethylene manufacture (Fig. 4.3)
- recycling of unreacted vinyl chloride monomer for feed to future batches in PVC manufacture (Fig. 4.6).

*Sustainable Process Engineering: Concepts, Strategies, Evaluation, and Implementation*
David Brennan
Copyright © 2013 Pan Stanford Publishing Pte. Ltd.
ISBN 978-981-4316-78-1 (Hardcover), 978-981-4364-22-5 (eBook)
www.panstanford.com

In these cases, unconverted reactants are returned to the reactor, leading to reductions in both raw material consumption and emissions to the environment. There are, however, other important considerations for recycling process streams. For example, in processes involving reactant recycle,

- *separation processes* are required to separate unconverted reactants from products leading to increased energy consumption
- recycled materials must often be *pumped* or *compressed* to reactor feed pressure contributing to increased energy consumption
- *material losses* may be incurred through imperfect separations and through purge streams occurring in the recycle loops contributing to waste emissions.

### 8.2.1  *Economics of Recycling Process Streams*

Recycling within processes enables savings in operating cost through

- reduction in raw materials consumption
- reduction in waste treatment or waste disposal.

However, it incurs additional capital and operating costs through possible

- additional separation processes
- compression or pumping of recycled materials.

Thus an economic balance between costs and benefits is incurred.

### 8.2.2  *Environmental Credits and Burdens of Recycling*

Environmental credits derive from reduced raw materials consumption and reduced emissions to the environment. Reduced raw materials consumption implies

- reduced resource depletion
- avoidance of environmental impacts in upstream extraction of raw materials and their subsequent processing.

Reduced emissions implies reduction in environmental impacts through various damage categories. *Environmental burdens* are incurred in recycling through materials separation and transport, mainly due to energy consumed.

Thus an *environmental balance* between benefits and burdens applies. Resolution of both economic and environmental balances is required to determine the viability of recycling.

## 8.3 Closed Loop and Open Loop Recycling

A distinction can be made between

- *closed loop* recycling and
- *open loop* recycling.

Figure 8.1 illustrates this distinction for a hypothetical reactor separator system.

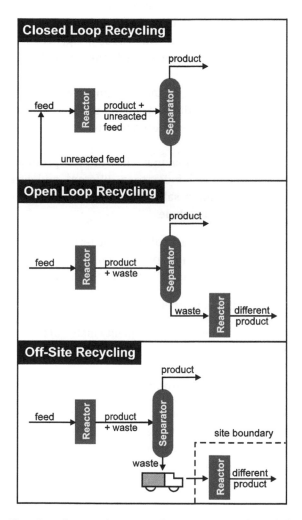

**Figure 8.1** Closed and open loop recycling options. Adapted from Allen and Rosselot (1997). Reproduced by permission of John Wiley and sons Inc.

*Closed loop recycling* within processes involves transfer of a material to an upstream position in the processing sequence. In product recycling, closed loop recycling involves transfer of material to an upstream position in the product's life cycle. Closed loop recycling may be practised with

(a) no material losses in the recycling loop
(b) material losses in the recycling loop.

Material losses are incurred when there is need to purge impurities to maintain a required concentration (or quality) of material in the recycle loop. This occurs, for example, in chlorine manufacture by brine electrolysis, where precipitated impurities from salt ($CaCO_3$, $Mg(OH)_2$) and any trace heavy metals are removed with some brine in the recirculated brine stream (see Fig. 4.1).

Material losses can also occur through imperfect separations in the recycling loop. For example, in the recycling of unreacted ethane to reactor feed (see Fig. 4.3) there are losses of ethane to other streams in the various separations by distillation.

*Open loop recycling* occurs when material from one production sequence is recovered, reprocessed, and fed into a different and often unrelated production sequence to make a saleable product. Examples of open loop recycling in a product context include

- recovery of post-consumer polyethylene terephthalate (PET) bottles which are then fed into fibre production;
- recovery of post-consumer PVC bottles for drain-pipe manufacture.

If waste is not recycled, it must either be

(a) treated to allow discharge of suitable quality material to the environment
(b) disposed of as landfill
(c) incinerated or
(d) discharged directly to the environment.

Within a product's life cycle, there is often a range of recycling options. Materials produced at different stages of a product's life cycle require different treatment processes as part of the recycling loop before material is returned to the product chain. Even for materials withdrawn from the same point in a product's life cycle chain, various options may exist in terms of

- treatment processes prior to recycling
- points of return for recycled material.

## 8.4  On-Site and Off-Site Recycling

A distinction may be made between *on-site* and *off-site* recycling. In *off-site* recycling (Fig. 8.1), waste is *transported* to another site where it is transformed to a saleable product. Transport of the waste by road, rail, ship, or pipeline incurs environmental impacts as well as public risk of containment loss.

### 8.4.1  *Examples of Off-Site Recycling*

Surplus hydrogen rich streams from catalytic reforming units in petroleum refineries are commonly piped off-site. Gases are separated from hydrocarbons with hydrogen bottled for supply to customers, while hydrocarbons are returned to the refinery or sent to a petrochemical plant (see Fig. 3.5).

Other common examples of off-site recycling include

- regeneration of spent catalysts for reuse
- regeneration of waste sulphuric acid from refinery alkylation units to produce saleable sulphuric acid
- recycling of used tyres, involving crumbing of spent rubber followed by re-blending into new tyre rubber
- waste oil recovery and treatment to produce saleable transport fuels.

## 8.5  Producer and Consumer Waste

For a product life cycle chain incorporating product manufacture, use and disposal, an important distinction exists between

- *producer waste* and
- *post-consumer waste.*

*Producer waste* is generated in *production* processes. Producer waste is usually relatively clean, free of contamination, and of known composition. Examples of producer waste include

- rejected materials such as glass from glass manufacture, paper from paper manufacture
- off-cuts from materials such as aluminium sheet, polyurethane foam.

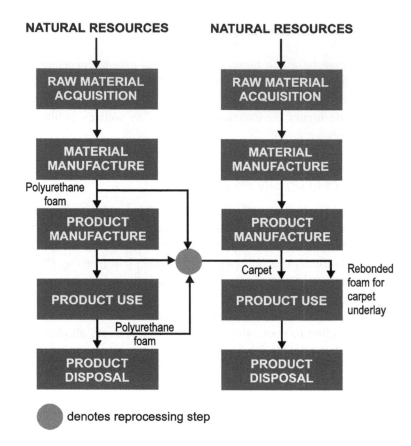

**Figure 8.2**  Open loop recycling options for polyurethane foam.

*Post-consumer waste* (also termed *municipal solid waste*; *MSW*) is generated by households, commercial establishments, and institutions. Mainly solid, its composition can vary depending on its source, and can change over time.

Figure 8.2 shows open loop recycling of producer and consumer polyurethane foam waste.

## 8.6  Hierarchical Approach to Materials Recycling

A hierarchical approach can be adopted towards recycling incorporating reduction, reuse, recycling, and recovery.

*Reduction* involves improving process and product design to make products with consistent properties, but incorporating less material. This has also been termed 'dematerialisation'. One example of reduction is the 375 ml glass beer bottle produced by an Australian glass manufacturer which in 1986 weighed 260 g in 1986 and in 1997 weighed 180 g. This 'light-weighting' leads to decreased raw material and energy consumption for virgin glass bottle production.

*Reuse* involves using same objects or materials again without changing their physical form. Examples include reuse of

- pallets, crockery, refillable bottles
- platinum catalysts, ion exchange resins, following regeneration.

*Recycling* involves use of waste material to reduce consumption of virgin raw materials. An example is the recycling of waste paper for paper board.

*Recovery* is usually related to energy recovery, but includes recovery of waste oils for heating and transport fuels.

Other terms sometimes used include a 'treat and dispose of' category, with the expectation of some residual waste requiring disposal, for example, by landfill.

## 8.7 Plastics Recycling

Plastic products have some environmental advantages, for example, their

- preserving facility reduces food wastage in storage, handling, and transportation
- low weight reduces transport impacts
- moulding capability reduces generation of producer waste.

About 35% of plastics are typically used for packaging. Packaging has a short life, with early disposal by consumer, often within weeks of its manufacture. Most plastics packaging has traditionally been disposed of at landfill sites. The chemical stability of plastics has contributed to the stability of landfill sites but has also contributed to the need to find more new sites.

Some examples of plastics recycling options are shown in Fig. 8.3. Note the optional points for withdrawal and re-entry to the main processing chain, as well as treatment options. Options include

(a) Recycling around the user stage (or reuse). An example is the returnable bottle system. Energy required is small compared with that for upstream processing to make new bottles.

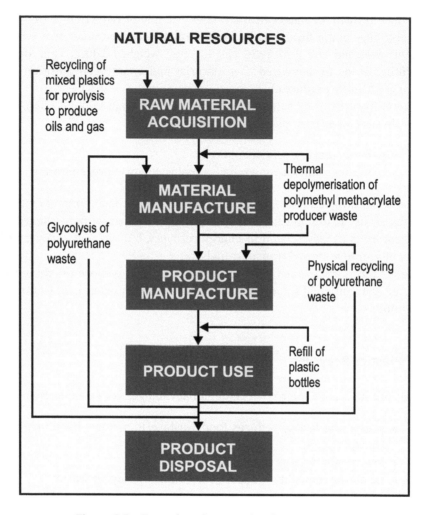

**Figure 8.3** Examples of options for plastics recycling.

(b) Material after product use is recycled to the raw materials acquisition stage. In this case the recycling loop would include

- recovery and cleaning of polymer
- chemical conversion of the polymer to an acceptable monomer or raw material form.

The recycling energy consumed in option (b) is likely to be increased beyond option (a) but there is a reduction in raw materials and energy consumption in the upstream processing needed for monomer production.

Some technologies for open and closed loop recycling of plastics include

**A. reuse in lower grade products as**

- inert filler, using chemical stability property
- polyurethane scrap used as carpet underlay

**B. secondary recycling with thermal reprocessing,** used with

- vinyl polymers
- condensation polymers
- polyalkenes

**C. chemical depolymerisation (or chemolysis)** used with

- polyurethanes
- polyethylene terephthalate (PET)

**D. thermal depolymerisation** used with poly-methyl methacrylate

**E. feedstock for hydrogenation, gasification, pyrolysis of hydrocarbons.**

This option has been of interest for mixed plastics from MSW.

Useful references on plastics recycling technology include those of Azapagic *et al.* (2003), Bhat (2007), Ehrig (1992), and Pongracz (2007) listed under supplementary reading.

## 8.8  Glass Recycling

Used and recovered glass (post-consumer or commercial), as well as off-specification glass, which is suitable for remelting is referred to as 'cullet'. Cullet is not only made into new containers, but also secondary markets such as fibreglass and glasphalt (a paving asphalt using crushed cullet to replace stone aggregate).

Cullet is one of four principal ingredients in container glass, along with sand, limestone, and soda ash. By melting at a lower temperature than glass-forming raw materials, cullet allows for a reduction of energy input to melting furnaces with an added benefit of prolonging refractory life.

Cullet for container production is a mixture of in-house rejects and post-consumer glass that has been beneficiated to remove contaminants. Glass manufacturers have compared the cost of batch raw materials to produce 1 ton of glass with the purchased price of cullet for recycling. For container glass, a 10% increase in cullet reduces melting energy by 2.5%, and emissions of particulates by 8%, of $NO_x$ by 4% and of $SO_x$ by 10%.

Several classification and separation issues, as well as economic constraints must, however, be addressed.

- Chemical composition differences exist between the largest product categories — glass containers, window and automotive glass, electronic glasses, fibreglass, and home cookware.
- Glass containers comprise approximately 4% by weight of MSW.
- Frequently only container and flat post consumer glass is recycled commercially.

## 8.9  Recycling of Materials from Products

Many industrial products comprise multiple materials. In order to recycle these component materials within the industrial economy, they must first be separated before being fed into the various recycle loops with their associated treatment requirements.

Examples of such products include motor vehicles, personal computers, and lead–acid batteries. Lead–acid batteries require separation of lead, dilute sulphuric acid ($\sim$10% concentration by weight), and polypropylene, and converting each component into reusable material.

An important consideration in both materials and product design is the ease of disassembly of products into component parts and the consequent ease of materials recycling. There can be conflicts between objectives of design for product function and product recyclability.

## 8.10  Waste Treatment Option

If waste is treated, costs are incurred and there are secondary environmental impacts. Raw materials and energy are consumed, and new wastes are generated.

If a marketable product results,

- treatment costs may be offset by product sales revenue
- environmental impacts resulting from waste treatment may be offset by impacts avoided in making the same product by an alternative method. For example, impacts in scrubbing waste $SO_2$ to make synthetic gypsum may offset impacts incurred in mining natural gypsum.

If waste is sent to landfill or incinerated, there are both

- associated charges and
- environmental impacts.

Some waste streams may contain single pollutants such as sulphur dioxide or ammonia within gas streams vented to atmosphere, and pose relatively straightforward tasks in processing to convert the pollutant to a marketable compound or a relatively harmless compound for disposal. However, many waste materials, particularly those remaining from the processing of minerals, are complex and represent considerable problems in relation to processing, marketability or disposal.

One example of such a waste is red mud from alumina refineries; red mud from MAL Aluminium Production's damaged tailings reservoir in Ajka, Hungary was reported (tce, 2010) to have a pH in the range 9.5 to 12 and contain 40%–45% iron oxide, 10%–15% aluminium oxide, 10%–15% silicon dioxide, 6%–10% calcium oxide, 5%–6% oxygen-bonded sodium oxide, and 4%–5% titanium dioxide. An important aspect of managing red mud waste from alumina refineries is the dewatering of the red mud discharged from the alumina refining process (Boger, 2010). Water is removed and recycled for further use, while the thickened mud can be pumped to an area for dry stacking. Application of rheological principles is a key part of the dewatering process (Boger, 2010).

## 8.11 Aqueous Effluent Treatment and Water Recycling

Water is a key input into many processes as a raw material, and into the generation of various energy related utilities used in processes. We can identify the use of water as a key raw material or agent material in the processes considered in Chapter 4. Examples include

- the production of chlorine-caustic soda involves use of water in the soda cell to produce hydrogen and caustic soda
- the production of sulphuric acid involves the use of water in the absorption of sulphur trioxide to produce sulphuric acid
- the production of ethylene uses water firstly as diluent steam in the ethane cracking furnace, and secondly as water in the quenching of reaction products after the waste heat boiler
- the batch production of PVC from vinyl chloride monomer uses water for suspending VCM droplets and PVC particles as they are formed.

The use of water in utility systems within a broad context was identified in Fig. 6.1, and in more detail through the various utility systems discussed in Chapter 6. Water also has a key role in many separation processes, for example, through use of live steam in distillation, aqueous solvents in gas absorption, and solids washing in filtration operations. Water is also used in equipment cleaning, either in direct cleaning or in rinsing following the use of cleaning agents.

The many usages of water within process and utility systems lead to a number of aqueous effluents of various compositions. Traditionally, these effluents have been collected and combined into a single stream for treatment, prior to discharge to a sewer. Some effluents may be suitable for direct discharge to sewer and can bypass the treatment.

Processes for treating aqueous effluents have traditionally been subdivided into

- primary
- secondary and
- tertiary treatment.

The objectives of *primary treatment* are to recover water as well as valuable chemicals, and prepare the aqueous effluents from primary treatment for secondary treatment. Primary treatment can involve many different separation steps, depending on the nature of contaminants or materials for recovery. Various solids removal processes are common, but other separation processes such as stripping with air or steam, membrane separation, ion exchange, and adsorption are widely used.

*Secondary treatment* involves biological processes, commonly aerobic and anaerobic treatment, to break down organic matter into stable compounds.

*Tertiary treatment* involves the final polishing of aqueous waste water for discharge and may involve nutrient (N and P) removal, ultrafiltration, and disinfection. Useful discussions of water treatment methodology include Zinkus, Byers, and Doerr (1998), Eckenfelder (2000), and Smith (2005).

Depending on the availability and cost of fresh water, the value of contaminants in the aqueous effluents, and the overall scale and complexity of the water system, opportunities exist for recycling or directly reusing certain effluent streams within the process. This approach recognises that not all water streams used as process or utility inputs need to be of the same quality. Design strategies for maximising water reuse based on mass flow rate and impurity concentration (single contaminant) of effluents are outlined by Smith (2005).

## 8.12 Disposal of Wastes

Traditionally, the most commonly encountered methods of waste disposal have been landfill and incineration. Specialised waste such as radioactive wastes from nuclear reactors, and carbon dioxide from power station flue gases, require specialised approaches.

### 8.12.1 *Landfill*

Landfills have traditionally been sited on the outskirts of communities, providing isolation from the health and environmental problems of decomposing wastes, such as odour, litter, and vermin. A modern sanitary landfill is an engineered site, selected, designed, and operated to minimise environmental impacts.

Municipal wastes are deposited in a confined area, spread in thin layers, compacted, and covered at the end of each working day. Key elements are:

- enclosing waste deposits underneath to prevent groundwater contamination
- covering waste deposits to limit emissions to air and rain water entry
- long-term stability to subsidence, erosion, and geological changes.

There is need to drain any liquid which collects within the waste and becomes contaminated. This drained liquid, *leachate*, requires processing before disposal; some concerns include pH and heavy metals content.

An important issue is *landfill gas*, generated by the degradation of organic materials. On modern sites, the gas is collected and flared, or preferably used as a low calorific value fuel, sometimes with electric power generation. In Melbourne, some landfill sites have associated electricity generation (e.g., 11 MW were being generated at Clayton, Victoria around 2000). Figure 8.4 shows some key features of a municipal landfill.

Alternatives to landfills are

- deep wells (used in the United States for disposal of hazardous liquid wastes) and
- old mines (e.g., salt mines in Germany and Poland). Salt mines offer very stable geological structure.

The number of landfills is generally decreasing (the number in the United States halved between 1988 and 1998), but individual landsites are becoming larger. Related factors include the following:

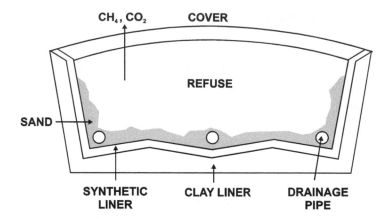

**Figure 8.4**  Schematic outline of a landfill with liners and leachate collection.

- Many old landfills, not meeting current standards, are being closed.
- It is becoming increasingly difficult to site new landfills, due to restricted availability of suitable land and to community opposition.
- Costs of site design, construction and operation, including leachate and gas collection and treatment, favour large scale facilities.

New landfill sites are becoming harder to establish because of public opposition. Objections are nuisance issues (dust, noise, traffic, odour), environmental concerns (groundwater pollution from leachate, migration of landfill gases to nearby properties), and land usage. Landfill costs are increasing due to more stringent site controls, promoted by legislation and community expectations. In some countries landfill taxes have been introduced.

Distinction should be made between

- municipal waste landfills and
- prescribed waste landfills.

A prescribed waste landfill might handle all of putrescible waste (plant and animal residues which are degraded by bacterial action), solid inert waste, low level contaminated soil, and prescribed wastes, but exclude soluble chemical wastes, hazardous wastes and liquid wastes.

### 8.12.2 *Incineration*

An alternative to landfill is incineration. Depending on the amount of $O_2$ present, treatment can be classified as

- pyrolysis or gasification involving starved air, or
- incineration involving direct combustion.

Temperature, residence time, and turbulence (3 T's) are important. Most incinerators operate with generous design allowances to ensure all combustible components are decomposed and converted to $CO_2$, $SO_2$, $H_2O$, HCl, $Cl_2$, $P_2O_5$, $N_2$. Temperatures and construction materials are important in downstream gas cleaning systems and stacks.

The main *EC directive* maintains an integrated pollution control approach. The directive sets out operating conditions and emission limits for hazardous waste incineration plants. Permits require that composition and characteristics of waste are made known to the incinerator operator. Combustion chamber temperature is maintained at 850°C for 2 seconds in the presence of 6% oxygen. If waste has >1% halogenated organics, temperature must exceed 1100°C. Emission limits for gases are specified in EC directive. Wastewater and waste heat release are also required to be minimised. Slow cooling of combustion gases in temperature range 450–200°C has been identified as a major source of dioxins (polychlorinated dibenzofurans and polychlorinated dibenzo-p-dioxins). Dioxins are highly toxic substances. Concentration limits for dioxins in emitted gases have been set at 1 ng/m$^3$, with a target of 0.1 ng/m$^3$.

A recent area of interest is the use of cement kilns (which operate at high temperatures) for combustion of waste car tyres and plastics. The application of incineration and energy recovery to MSW is discussed by Kirkby and Azapagic (2004).

## 8.13 Concluding Remarks

Materials recycling can occur within processes and within the product life cycle. In the waste management hierarchy (discussed in Chapter 2) recycling is preferred to waste treatment which in turn is preferred to waste disposal.

Recycling has economic and environmental benefits but it may also incur economic and environmental burdens. It is important to identify, evaluate, and resolve all benefits and burdens in decision-making.

It is easier to recycle materials of single composition than composite materials of complex composition if it is desired to recover individual components of the composite material. This is an important consideration in selecting materials for products. An example is materials selection for body, engine, and internal cabin components in motor vehicles. Product design for ease of disassembly may also be important in certain cases.

Offsite recycling may be seen to come under the umbrella of industrial ecology, both for processes and for products. In future generations, materials recycling may become of even greater strategic importance as a source of raw materials if certain metallic ores become scarcer, or of substantially lower grades.

## References

Allen, D. T. and Rosselot, K. S. (1997) *Pollution Prevention for Chemical Processes*, John Wiley and Sons, New York.

Boger, D. (2010) Back to basics, *The Chemical Engineer*, 834/5, 28–29.

Eckenfelder, W. (2000) *Industrial Water Pollution Control*, 3rd edn, McGraw-Hill, New York.

Kirkby, N., and Azapagic, A. (2004) Municipal solid waste management, in *Sustainable Development in Practice* (eds Azapagic, A., Perdan, S., and Clift., R.), Wiley, Chichester, England.

Smith, R. (2005) *Chemical Process Design and Integration*, John Wiley and Sons tce NEWS (2010) Hungarian red mud contained, *The Chemical Engineer*, November, 10.

Zinkus, G. A., Byers, W. D., and Doerr, W. W. (1998) Identify appropriate water reclamation technologies, *Chem. Eng. Prog.*, May, 19–31.

## Supplementary Reading

The following references provide additional reading material related to plastics recycling and packaging:

Azapagic, A., Emsley, A., and Hamerton, I. (2003) *Polymers: The Environment and Sustainable Development*, John Wiley and Sons, Chichester, England.

Bhat, G. (2007) *Processing Postconsumer Recycled Plastics*, John Wiley and Sons, Hoboken, NJ, Chapter 12, pp. 357–383).

Ehrig, R. J. (ed.) (1992) *Plastics Recycling: Products and Processes*, Hanser, New York.

Pongracz, E. (2007) The Environmental impacts of packaging (Chapter 9), in *Handbook of Environmentally Conscious Materials and Chemical Processing* (ed. Kutz, M.), John Wiley and Sons, Hoboken, NJ, pp. 237–278.

# Chapter 9

# Waste Minimisation in Operations

## 9.1 Non-Flow-Sheet Emissions from a Process Plant

Some process effluents, such as purges and terminal streams to atmosphere and to drain, are identifiable from a well-defined process flow sheet supported by mass and energy balances. These effluents reflect steady state operation of the plant according to design intent.

However, there are other sources of effluents possible, derived both from the physical plant and the complexities of its operating procedures. These sources include

- plant start-up and shut-down procedures
- abnormal operating conditions of the plant
- maintenance of equipment and plant
- cleaning and purging of equipment and plant
- product grade changes in multiproduct plants necessary to achieve distinct properties for different product applications
- defects in equipment and piping.

For many of these cases, pressures, temperatures, flow rates, and compositions of process streams can differ from design intent. Special provisions must be made for all of these cases to protect the environment.

Process control has an important role in minimising waste in operations. This role is exercised through monitoring process conditions using a comprehensive range of instrumentation, and adjusting to deviations from

*Sustainable Process Engineering: Concepts, Strategies, Evaluation, and Implementation*
David Brennan
Copyright © 2013 Pan Stanford Publishing Pte. Ltd.
ISBN 978-981-4316-78-1 (Hardcover), 978-981-4364-22-5 (eBook)
www.panstanford.com

intended conditions using control systems. Analytical instruments are finding increased use for monitoring concentrations of different molecular species in process and effluent streams, in ambient air and water bodies, and wastes at various stages of treatment and disposal. Good process control systems, as well as accounting for deviations from intended process conditions, can also assist smoother start up and shut-down routines.

Despite good process control systems, failures in equipment, instrumentation, piping and valves, and utility supply, as well as human error occur periodically, and can lead to major environmental accidents.

## 9.2 Plant Start-Up

During start-up, it is necessary to bring temperatures, pressures, and concentrations of process streams to design conditions to ensure reaction and separation steps perform satisfactorily. Even after the main preparatory steps have been made to achieve required process conditions, a period is necessary to approach a stable operating regime. The duration of this period depends on the nature of the process, its inherent stability and complexity, and the supporting process control system. During this approach period, it is difficult to avoid additional wastes being formed.

One important task during start-up is the purging of equipment and piping to remove air before introducing flammable process materials. Nitrogen is most commonly used for this purpose. Once the atmosphere in the process equipment has been rendered inert, the reactant feed material can be introduced. There is a period, however, where the nitrogen/process material mixture must be purged from the plant safely. This may involve combustion or some other disposal process which incurs additional environmental impact.

### 9.2.1 *Case Example — Starting Up a Sulphuric Acid Plant (Refer to Case 4 in Chapter 4)*

Catalyst temperatures in the reactor need to be raised to approximately 430°C for required equilibrium and kinetic conditions for the oxidation of sulphur dioxide to sulphur trioxide. For a sulphur burning plant this can be done by burning fuel in dry air upstream of the reactor, and passing the hot air through the catalyst beds. However, it is difficult to achieve

- steady state conditions due to heat losses
- smooth transition from fuel burning mode to sulphur burning mode.

Hence when the transition is made,

- catalyst temperatures may not be optimal for conversion of $SO_2$ to $SO_3$
- gas and acid temperatures and acid concentrations may not be optimal for $SO_3$ absorption.

This leads to potentially higher emissions of $SO_2$ and $SO_3$ during start-up than would be encountered under normal steady-state operation of the plant.

## 9.3  Shut-Down of a Plant

It is important to make a number of provisions to minimise waste at plant shut-downs. The extent of the provisions depends on the duration and purpose of the shut-down. Shut-downs can be emergency or routine. Examples of essential provisions include

- drainage of equipment (e.g., reactors, columns) to suitable tanks
- provision of space at the base of vapour–liquid contacting columns for collecting drainage from trays or packings
- pressure venting from reactors and vessels
- pipeline drainage and purging of liquids that solidify on cooling
- purging of gas lines using inert gas
- related allowances in instrumentation and process control.

For shut-down, purging toxic or flammable process material from the plant may be necessary. In such cases, purging and venting must be directed to suitable sinks, such as flares or scrubbing towers.

## 9.4  Abnormal Operation

Abnormal operation covers a wide range of deviations from design intent. Possible deviations could include

- faulty opening of a relief valve or bursting disc
- membrane rupture
- failure of a utility supply, for example, electricity or cooling water
- variation in reactor feed composition
- failure of a valve
- failure of instrument or process control system
- human operator error
- corrosion, erosion, or mechanical failure in piping or equipment.

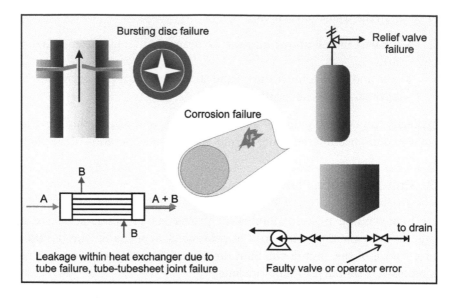

**Figure 9.1**   Examples of deviation from intended behaviour.

These deviations are similar to those explored in a hazard and operability (HAZOP) study, where the primary consideration is safety (see Chapter 12). However, environmental consequences of many of these deviations can be serious and must be evaluated. Some examples of abnormal events are shown in Fig. 9.1. In some cases, fluids can be released directly to atmosphere, waterways, or land. In other cases, process or utility streams can become contaminated, leading to generation of product waste, or other releases to the environment.

Equipment or piping failure can lead to leakage of fluid into the environment. This could occur

- directly, for example, at a pipe flange, or
- indirectly, for example, by leakage in a heat exchanger into a separate process stream or into cooling water.

## 9.5  Plant Maintenance

When equipment is removed from operational service for maintenance it must be opened up for cleaning, inspection and repair. For example,

- a worn pump impeller may need to be replaced
- a statutory pressure vessel inspection may be required

- a heat exchanger, or pipe flange, may have developed a leak needing to be sealed.

In isolating an equipment item from operating service for maintenance purposes:

- the draining or venting of process fluids may be a source of emissions
- if the plant continues to operate, process abnormalities may be caused by the absence of the specific equipment item from the process
- complete plant shut-down may be necessary with its implications for emissions.

## 9.6  Cleaning of Plant and Equipment

When a vessel, filter, or heat exchanger is cleaned, the deposits may be difficult to remove. The deposits themselves constitute an accumulated waste.

Water, steam, acid, or alkaline cleaning may be effective in removing deposits, but the resulting slurries or liquids may in turn need purification, due to contamination by the deposit material. If acid or alkali cleaning is used, subsequent neutralisation or rinsing may be needed, thus causing yet another waste — the product of the neutralisation or rinsing step.

It is important to

- understand the fouling mechanism and take preventive action
- develop a cleaning method which minimises waste.

## 9.7  Fouling

Fouling involves the formation of an undesired deposit on the surface of equipment, equipment internals, or piping during plant operations. Some common causes of fouling include

- particulate matter in fluid adheres to a surface (e.g., heat transfer surface)
- deposition of solids due to evaporation or solubility changes
- polymerisation

- decomposition (e.g., carbon formation)
- corrosion
- biological growth
- low cooling water velocity in heat exchanger tubes.

**Common effects of fouling are**

- reduced rate of heat transfer
- increased pressure drop.

**To minimise fouling**

- gain an understanding of the composition and properties of a deposit and its means of formation
- monitor temperatures of process streams and equipment surfaces, and maintain them within allowable limits
- monitor velocities of process and utility streams, and maintain them within allowable limits
- monitor compositions of process streams, and where possible avoid contaminants which promote fouling.

## 9.8  Transport and Storage of Raw Materials and Products

Environmental burdens and risks in transport and storage of process materials were identified in Chapter 3 (Section 3.3) as a major driving force for improved industrial ecology. Figure 9.2 summarises the essential steps

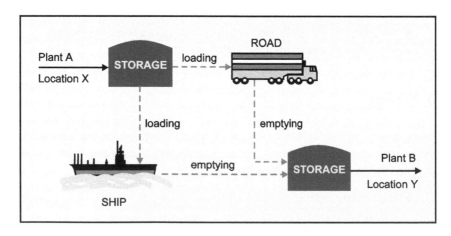

**Figure 9.2**  Transport of product from Plant A to Plant B.

in transport of process materials incurring storage, loading, and unloading activities, as well as the transport operation. Road and rail transport of hazardous materials has serious potential for damage to the environment and to humans, with some serious incidents recorded in the past, involving multiple fatalities of civilians. However, there have also been many environmental disasters from marine transport, including instances of lost lives. A well known, but by no means *only*, example is the crude oil spill from the Exxon Valdez. Marine transport of bulk materials transport is discussed by Rawson (1994).

Environmental damage from spillages in marine transport is of particular concern when spillages occur at environmentally sensitive locations such as rivers, near coastlines, and areas of major environmental significance such as the Great Barrier Reef, off the north east coast of Australia. It is important to recognise that fuel oil leaks can occur even with benign cargoes.

Environmental risks resulting from accidental release of materials from storage, transport, and handling, as well as processing, have often been omitted from life cycle assessment studies performed thus far. In the event of a spill, containment of a liquid is more difficult than containment of a solid. This is one reason why bulk transport of liquid sodium cyanide is restricted. Crude oil spills are a good example of the difficulty in containing liquid spills.

An important approach to minimising environmental impacts from transport and storage of materials lies in effective industrial planning and site selection of process plants. This is discussed in Chapter 15. In operations, emphasis should be on well-defined operating, maintenance, and emergency response procedures. Careful consideration of transport routes, integrity of transport vehicles, and personnel training is also important.

### 9.8.1 *Storage Tanks*

Frequent environmental impacts from storage tanks include

- periodic removal of sludge from tank bottoms
- emissions to air of volatile organic compounds (VOCs) from roof vents on atmospheric pressure tanks.

Sludge comprises corrosion products and heavy suspended matter from material being stored. If impurities are compatible with downstream use, they may be suspended by mixing or emulsifying techniques. Otherwise, they must be periodically removed and either treated or disposed of.

VOC emissions result from increase in liquid levels in tanks storing VOCs. Level increases are caused by

- filling operations (referred to as 'working losses') which result in vapour displacement
- changes in atmospheric temperature or pressure (referred to as 'standing losses'), for example, expansion of liquid contents due to increase in ambient temperatures.

Storage tank emissions are influenced by tank design which may be

- fixed roof, of domed or conical design
- internal or external floating roof design.

Emission quantities depend on the vapour pressure of the stored liquid, tank design, paint colour and condition, and tank location. Fixed roof storage tanks are cheapest to construct but have larger emissions. Treatment of released vapours externally from the tank (e.g., by adsorption, liquid scrubbing) can be effective in reducing emissions but incurs additional costs.

In the case of floating roofs there are several design options:

- An external floating roof tank without a fixed roof, where a floating deck rises and falls with change in liquid level. Emissions result from evaporation of liquid from the exposed shell wall.
- A domed external roof tank, similar to external floating roof design but with a self-supported domed roof above the floating roof. This design of floating roof tank is claimed to achieve a 60% to 99% reduction in emissions compared to a fixed roof tank.

Details of tank design and construction may be found in

- Perry's *Chemical Engineers' Handbook*
- US EPA website under www.epa.gov/ttn/chief which also provides estimating procedures and software (TANKS program) for estimating emissions.

### 9.8.2 *Major Environmental Incidents Arising from Storage*

Major environmental damage can occur in the event of tank failure, derived from the large inventory of material stored. Similar approaches to HAZOP studies can be adopted to examine possible causes and consequences of failure. Special design approaches can be employed to minimise damage in the event of spillages or releases to atmosphere, for example,

- bunds to contain spilled liquid
- tank location and diversion schemes to prevent contamination of rivers or groundwater

- containment of storage within ventilated buildings where toxic materials are being stored.

## 9.9  Fugitive Emissions

Fugitive emissions are unintentional releases of process fluid from equipment or open areas such as drains. Any plant component prone to leak is a potential source of fugitive emissions. Plant components contributing to fugitive emissions include valves (including pressure relief valves), pump and compressor seals, and pipe flanges.

Emission rates from a plant component can be estimated from the relationship

$$E = m_{\mathrm{voc}} \times f_{\mathrm{av}}$$

where

$E$ = emission rate (say of VOCs)

$m_{\mathrm{voc}}$ = mass fraction of VOC in process stream undergoing leakage

$f_{\mathrm{av}}$ = average emission factor for a plant component.

Some estimates of emission factors for combinations of plant item and process fluid are shown in Table 9.1. Table 9.2 lists typical numbers of emission sources for selected refinery units. Data for Table 9.2 may not be an accurate reflection of current plant designs for refinery units, but illustrates the accounting approach.

Table 9.1  Average emission factors for estimating refinery fugitive emissions. Extracted from Allen & Rosselot (1997) p. 135

| Equipment | Service | Emission factor, $f_{\mathrm{av}}$ kg/hr/source |
|---|---|---|
| *Valves* | Hydrocarbon gas | 0.027 |
| | Light liquid | 0.011 |
| | Heavy liquid | 0.0002 |
| | Hydrogen gas | 0.0083 |
| *Pump seals* | Light liquid | 0.11 |
| | Heavy liquid | 0.021 |
| *Compressor seals* | Hydrocarbon gas | 0.63 |
| | Hydrogen gas | 0.05 |
| *Pressure relief valves* | Hydrocarbon gas | 0.16 |
| | Liquid | 0.007 |
| *Flanges, other connectors* | All | 0.00025 |
| *Open-ended lines* | All | 0.002 |

Table 9.2   Estimates of number of individual emission sources for selected refinery units (US EPA, 1980)

| Process unit | Valves | Flanges | Pump seals | Compressor seals | Drains | Relief valves |
|---|---|---|---|---|---|---|
| Catalytic hydrotreating | 650 | 2600 | 10 | 3 | 24 | 6 |
| Catalytic reforming | 690 | 2760 | 14 | 3 | 49 | 6 |
| Hydrogen production | 180 | 640 | 5 | 3 | 17 | 4 |
| Sulphur recovery | 200 | 800 | 6 | 0 | 20 | 4 |

More refined techniques for estimating fugitive emissions on an operating plant rely on source specific emission tests or continuous emission monitors. Fugitive emissions may be minimised by attention to design, monitoring, and maintenance of equipment and piping. For a proposed plant, including proposed modifications to existing plant, systematic identification of potential sources of fugitive emissions, as well as accidental spills, requires a well-developed piping and instrumentation diagram (see Chapter 16).

## 9.10   Environmental Risks Resulting from Storm Water

Storm water can entrain solids and immiscible liquids such as oils, as well as dissolve soluble salts. Such salts may be present as spillage or may accumulate as dust on equipment or piping surfaces. An example of a plant where storm water collection and treatment is critical is solid sodium cyanide manufacture; in this case the cyanide salt is toxic.

Proper provisions must be made for collection, assessment, and treatment of storm water from all process plants prior to discharge to the environment. Collection provisions should reflect rain intensity, which varies during the year, and can be high in tropical regions. For further reading see Garg and Pair (1995).

## 9.11   Risks in Mining and Extraction of Materials

In extending our focus beyond process plants and the storage and distribution of raw materials and products, there are important environmental and safety risks in the mining and extraction of raw materials, minerals, and fuels.

Examples include the risks of explosions in coal mines, of contamination of land and waterways in mining and mineral processing, and of oil leaks in the exploration and extraction of petroleum. A recent example in oil extraction (April 2010) is the explosion and fire on the Deepwater Horizon drilling rig and subsequent oil leak in the Gulf of Mexico threatening the coast of Louisiana, Texas and the Mississippi estuary in United States. Estimates of leakage rates reported in news media ranged from 1000 to 8000 barrels a day.

## 9.12  Concluding Remarks

We have seen how it is necessary to look beyond the design case to the operating plant, beyond intended plant operation to abnormal plant operation, beyond plant operation to plant maintenance and cleaning, and beyond the boundaries of process plants to transport and storage of raw materials and products. This approach is similar to the progressive consideration of detail in the design of a process plant design, from process to equipment to plant, and from process flow sheet to piping and instrumentation diagram.

Failure to contain spillages during operation can result in long term damage to land and groundwater. Such damage can have major implications following plant closure and decommissioning. Site remediation can be lengthy, technologically difficult, and costly, and is not uncommon on processing sites which have been in use over many decades. One example is the Botany chemical site in Australia, where Orica are undertaking a number of remediation initiatives (see http://www.oricabotanytransformation.com/.) This emphasises the importance of provision for collection and containment of spillage, as well as high standards in operations, maintenance, and cleaning of plant and equipment.

The contribution of fugitive emissions to total emissions from a process plant and site can be major. Fugitive emissions can be minimised in design both by limiting the number of potential emission sources and by careful selection of seals, valves and fittings. Minimisation of fugitive emissions in operations can be assisted by systematically monitoring plant emissions, and regular inspection and maintenance of components contributing to fugitive emissions.

## References

Allen, D. T. and Rosselot, K. S. (1997) Pollution Prevention for Chemical Processes, John Wiley and Sons, New York.

Garg, D. and Pair, R. B., Jr. (1995) Effectively manage stormwater in a CPI complex, *Chem. Eng. Prog.*, 91(5), 70–76.

Rawson, K. ed. (1994) The Carriage of Bulk Oil and Chemicals at Sea, Institution of Chemical Engineers.

US EPA. (1980) *Assessment of Atmospheric emissions from Petroleum Refining*, EPA-600/2-80-075a, Environmental Protection Agency, Research Triangle Park, NC.

# Problems: Part B

## 1. Chlorine Production and Purification

Figure 4.1 showed a simplified flow diagram for the reaction and feed systems of a mercury cell plant for producing chlorine and caustic soda.

(a) In the operation of such a plant identify the

   (i)  potential sources of mercury release to the environment and
   (ii) potential contamination of product streams with mercury.

   Postulate the potential damage which could result from such releases and product contamination.

(b) With reference to Fig. 4.2 explain the essential features of a membrane cell for chlor-alkali production. What is the key advantage of the membrane cell compare with the mercury cell? Are there any environmental disadvantages of the membrane cell?

(c) Figure 5.4 showed a simplified flow diagram for the purification of chlorine from electrolytic cells. It is proposed to cool and dry chlorine gas generated in a membrane cell prior to compression, for piping to a customer plant. Chlorine is to be dried using concentrated sulphuric acid. Recommend strategies for

   (i)  minimising consumption of sulphuric acid in the drying operation
   (ii) recycling dilute acid for use within the drying plant.

(d) It is proposed to compress chlorine gas from a 200 tonnes per day chlorine plant. Chlorine will be compressed from suction conditions of 1 bar abs, 35°C to a delivery pressure of 3 bar abs.

   (i)  Assuming adiabatic compression, estimate by calculation the power requirements for the compressor.
   (ii) Calculate the temperature of chlorine gas leaving the compressor.
   (iii) If the chlorine gas leaving the compressor were cooled to 40°C using cooling water available at 28°C, calculate the required cooling water

consumption stating the assumptions made. What advantage might air cooling have over water cooling for this duty?

**Properties of Chlorine Gas**
Density at 1 bar abs, $0°C = 3.22$ kg/m$^3$
Ratio of specific heats $\gamma = 1.36$
Specific heat of chlorine gas $= 0.5$ kJ/kg K
Specific heat of water $= 4.2$ kJ/kg K

## 2. PVC Production from Vinyl Chloride Monomer

A simplified flow sheet and process description for the batch production of PVC from vinyl chloride monomer was provided in Chapter 4 with some supporting literature references.

(a) Why is cooling of the PVC reactor necessary, and by what means is it carried out in industrial practice?
(b) Why is conversion of VCM to PVC in the batch reactors incomplete?
(c) Why is refrigeration of the VCM vent stream released during slurry degassing necessary? Assume the vent stream composition is 97.2% VCM, 1.3% $H_2O$, 1.3% $N_2$, 0.2% $CO_2$ by mass.
(d) Why is incineration of the VCM vent stream released during slurry degassing necessary?
(e) Give two possible causes from within the PVC manufacturing process of producing 'out of specification' PVC product.

### Useful References

Ullmann (2002) *Encyclopaedia of Industrial Chemistry*, Wiley-VCH, Weinheim, Germany.
Websites for PVC manufacturers, for example:
European Vinyls Corporation: http://www.vinyl2010.org/ and http://www.vinylplus.eu/
Australian Vinyls: www.av.com.au
Vinnolit: http://www.vinnolit.de/vinnolit.nsf/id/EN_Home

## 3. Waste Minimisation in Nitric Acid Production

Nitric acid is manufactured commercially by the catalytic oxidation of ammonia to nitric oxide at elevated temperature, followed by low temperature oxidation of nitric oxide to nitrogen dioxide, and absorption of nitrogen dioxide in water (in the medium of a dilute nitric acid solution). Variations of the process exist especially in relation to selection of design pressure. Some useful references dealing with the process are appended below.

(a) Identify all process wastes arising from the manufacture of nitric acid from ammonia, assuming ammonia is supplied as a liquid stored under atmospheric pressure. Indicate the type of environmental damage caused by each waste. Identify the source of each waste, and explain how the waste is formed. Identify design and operational strategies to minimise the formation of each waste. Consider

   (i) normal operation
   (ii) start-up
   (iii) shut-down
   (iv) abnormal operation.

(b) Identify all potential utility wastes arising from the process, stating your assumptions. Also, identify any potential credits for avoiding waste derived from utilities generation within the process. Indicate the type of environmental damage, or credits relevant to each waste.

(c) For the process wastes identified in part (a), identify and explain concisely alternative treatment methods (end of pipe approaches). Provide a table summarising the environmental burdens and cost considerations for the alternative methods.

(d) Concern has been expressed about the effectiveness of the absorber in absorbing nitrogen dioxide gas ($NO_2$) in water. Suggest two changes in process conditions which might increase the recovery of $NO_2$ in the absorber.

(e) Consider a nitric acid plant where gases leaving the reactor are cooled from 900°C to 230°C to generate steam at 20 bar abs pressure and 330°C. Suggest how the steam would be *generated* and *used* within the nitric acid plant.

## Useful References to Nitric Acid Process Technology

Azapagic, A., Duff, C., and Clift, R. (2004) Integrated prevention and control of air pollution: the case of nitrogen oxides, in *Sustainable Development in Practice* (A. Azapagic, S. Perdan, and R. Clift., eds. ), Wiley, Chichester, UK.

Moulin, J. A., Makkee, M., and Van Diepen, A. (2001) *Chemical Process Technology*, Wiley, Chichester, UK.

Sinnott, R. and Towler, G. (2009) *Chemical Engineering Design*, Elsevier, Oxford, UK, Ch. 4.

Ullmann (2002) *Encyclopaedia of Industrial Chemistry*, Wiley-VCH, Weinheim, Germany.

## 4. Steam Generation

It is planned to generate 12 tonnes per hour of saturated steam by combustion of natural gas in a boiler. The steam will be generated at a pressure of 10 bar abs for use in a plant. Ten per cent excess air over that required for stoichiometric combustion will be used. Boiler efficiency, defined as the ratio of heat used in steam generation to heat released by fuel combustion, is 75%.

(a) Indicate, without making actual calculations,

    (i) how the flame temperature in the boiler furnace could be estimated by calculation

    (ii) why excess air is used in the fuel combustion

    (iii) the major source of the inefficiency in the boiler

    (iv) two common losses of steam from the steam circuit, and why they occur.

(b) Using the data provided below, calculate

    (i) the mass flow rate of natural gas consumed

    (ii) the mass flow rate and mass composition of the flue gas.

    Simplify the calculation by assuming that natural gas composition is pure methane, combustion air is dry, and there is negligible $NO_x$ in the flue gas.

(c) A source of 200 kg/h of hydrogen gas is available for burning in the boiler.

    (i) What reduction in the natural gas consumption can be achieved by burning the available hydrogen fuel?

    (ii) Quantify the benefits in environmental emissions from burning the supplementary hydrogen.

(d) Consider a case where saturated steam at 10 bar abs is used for heating within the plant. It is proposed to reduce the pressure of the resulting condensate from 10 bar abs to 1 bar abs pressure through a valve. Calculate the proportion of the condensate which flashes off as steam.

(e) It is proposed to modify the boiler to produce steam at 10 bar abs pressure but superheated to 300°C. If steam from the superheater were expanded in a perfect steam turbine to 1 bar abs pressure, determine the

temperature and condition of the steam at the turbine outlet. State your assumptions.
(f) Give concise details of the methods used to remove dissolved oxygen from boiler feed water in steam generation. Why is oxygen removal important?

Data for Question 4
1. Fuel properties
    Heat of combustion of methane:                    50 MJ/kg
    Heat of combustion of hydrogen:               142 MJ/kg
2. Steam properties

| | | **Saturated Water and Steam** | | | |
|---|---|---|---|---|---|
| P (bar abs) | T (°C) | hf (kJ/kg) | hg (kJ/kg) | sf (kJ/kg) | sg (kJ/kg) |
| 1 | 99.6 | 417 | 2675 | 1.303 | 7.359 |
| 10 | 179.9 | 763 | 2778 | 2.138 | 6.586 |
| | | **Superheated Steam** | | | |
| 10 | 300 | | 3052 | | 7.124 |

**Note:** h, enthalpy; s entropy; f, water; g, gas.

## 5. Recirculated Cooling Water

Water is supplied from an industrial cooling tower at a rate of 0.52 m$^3$/s to a process plant. The ambient wet bulb temperature is 23°C, the cooling water supply temperature is 27°C, and return temperature of cooling water to the tower is 49°C. The cooling water is delivered from the circulation pump(s) at a pressure of 6 bar g.

(a) Estimate (using calculation) the evaporation loss in the cooling tower.
(b) Assuming negligible entrainment loss from the cooling tower, estimate the make-up and purge mass flow rates in the circuit if make-up water contains 30 ppm total dissolved solids (TDS) and the recirculated cooling water contains 300 ppm TDS.
(c) Identify two possible aspects of water quality which might contribute to the 300 ppm limit on dissolved solids concentration referred to in part (b).
(d) Estimate the power required to circulate the cooling water, stating the assumptions made.

(e) Recommend the number of cooling water pumps and the drive type for each of these pumps. Give your reasons.

(f) Consider process heat exchangers in a plant located in your region where cooling is achieved using recirculated cooling water. Nominate typical *maximum inlet* and *maximum outlet* cooling water temperatures for heat exchangers operating in *summer*. What factors govern these two maximum temperatures you have nominated?

## 6. Materials Recycling

Select a particular material from

- glass containers
- aluminium
- newsprint
- lead
- polyurethane foam
- steel.

For the material selected

(a) Identify the processing, materials handling and transport steps associated with the recycling activity. Identify any changes in material composition during recycling.

(b) Develop a simplified life cycle model of the material flow, identifying its production, use and disposal phases. Establish the relationship of the recycling activity to the life cycle material flow; in particular, identify the points at which the recycled material leaves and re-enters the production flow chain.

(c) Identify key contributors to capital and operating costs for the recycling.

(d) List the environmental impacts

   (i)  saved by recycling

 (ii)  incurred in recycling.

(e) Make an assessment of the overall viability of the recycling case considered.

Aspects for consideration in the material recycling study:

   (i)  To what extent are material properties enhanced or degraded during recycling?

 (ii)  What is the economic value of the material being recycled?

(iii) What is the economic value of the resources used to make the recycled material?

(iv) How complex is the composition of the recycled material?

(v) What are the properties of the recycled material?

(vi) Where is the material withdrawn and returned in relation to the product life cycle, and what material transformations occur between these points?

(vii) What transportation distances are incurred during the recycling activity?

(viii) What are the limitations in available technology?

(ix) How favourable are the scale economies?

(x) What government incentives for recycling apply?

Note: Reference to Chapter 13 could assist with part (c).

# PART C

# EVALUATION

# Chapter 10

# Life Cycle Assessment

## 10.1 Introduction

In this chapter we explore the methodology of life cycle assessment (LCA) for use in the environmental assessment of processes and products.

LCA is a method which has been developed to provide a *quantitative* assessment of the environmental impact of a product over its *entire life cycle*. LCA accounts for both inputs and outputs, each derived from both process and utility sources:

- **inputs** — consumptions of raw materials, energy, fuel and water
- **outputs** — emissions to air, land and water.

Figure 10.1 depicts the key stages of a product life cycle with inputs and outputs incurred at each stage. In some cases, there may also be consumptions and emissions at the product disposal stage. By-products are grouped with emissions as outputs, since by-products are distinct from the main product processing and use chain. Market demand for by-products dictates whether by-products can be usefully sold into other product chains, or alternatively require disposal. The disposal option may involve additional prior treatment.

*Sustainable Process Engineering: Concepts, Strategies, Evaluation, and Implementation*
David Brennan
Copyright © 2013 Pan Stanford Publishing Pte. Ltd.
ISBN 978-981-4316-78-1 (Hardcover), 978-981-4364-22-5 (eBook)
www.panstanford.com

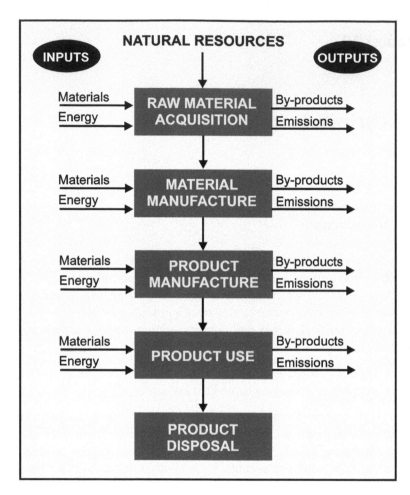

**Figure 10.1** Inputs and outputs for an LCA study (Brennan, 2007) [Brennan Ch. 3 in Kutz ed. 'Environmentally Conscious Materials and Chemicals Processing' © 2007 John Wiley and Sons. Reproduced with permission of John Wiley and Sons, Inc.]

## 10.2 Product and Process Applications

LCA has been most widely applied to products. One objective has been to establish which of several competing products has the least environmental impact for a given product function. Another objective has been to determine which part of a product's life cycle incurs the greatest environmental impact; this then directs activities such as sourcing alternative raw materials, developing an improved processing technology, or improving product design.

Product functions can be very diverse, for example,

- beverage containers
- paints for surface coating
- packaging for products
- automobiles.

LCA can be applied to competing materials for product applications, for example,

- steel, aluminium, or polymers for use in various automobile components
- alternative construction materials for buildings.

LCA can also be applied to processes in the context of process technology selection, process development, or process synthesis. A chemical processing plant is one module of an extended product life cycle which may involve more than one chemical processing stage. For example, both alumina refining and aluminium smelting are separate and significant processes in terms of environmental impact within the life cycle of the product aluminium. LCA is thus a valuable tool for quantitative evaluation of the 'cleanliness' of a process, and for ranking processes according to their environmental merits.

LCA can be applied to products and processes in various engineering and business contexts, for example,

- at research and development phase
- at the design stage
- in operations
- for competitive analysis in business review.

While elements of LCA thinking emerged in the 1960s, LCA became established as a methodology in the early 1990s. The methodology is still evolving and there are distinctions in the approaches, terminology, and definitions adopted so far. Not all aspects of the methodology have universal agreement. However, industries are increasingly adopting practices which incorporate LCA principles.

Some of the more accepted approaches to LCA methodology originate from

- Centre for Environmental Studies (CML), Leiden University in Holland
- Society of Environmental Toxicology and Chemistry (SETAC)
- United Nations Environment Program (UNEP).

LCA methodology is documented in international standards:

ISO 14040 (Principles and Framework)
ISO 14041 (Goal, Scope Definition and Inventory Analysis)
ISO 14042 (Life Cycle Impact Assessment)
ISO 14043 (Life Cycle Interpretation)

## 10.3 Basic Steps in Life Cycle Assessment

LCA has four basic steps:

- **Goal definition**, where the scope and purpose of the LCA is defined. The system boundaries are defined and the functional unit is established. The functional unit is the reference point to which environmental impacts are attributed.
- **Inventory analysis**, where all material and energy resources consumed and all wastes emitted are quantified. Note that *inventory* in LCA refers to the quantities of raw materials and energy inputs and emission outputs, not a quantity of stored chemical as the term denotes in safety assessment.
- **Classification**, where inventory data are grouped according to categories of specific environmental effects or impacts. The inventory data are then weighted within their respective impact category and aggregated to provide a numerical score for the impact potential. The environmental impact scores in each impact category are often *normalised* by dividing the impact score by the total impact score in that category resulting from all activities within a country. This procedure enables a comparison of the relative contributions by the product or process under study to a country's environmental impact for different impact categories. Alternatively, normalisation can be done on a regional or global basis.
- **Improvement analysis**, where aspects for improvement in processing or product use are identified and explored.

In earlier stages of LCA methodology development, an additional step, **evaluation**, often followed the classification step. Evaluation involves weighting the environmental impact scores in the various impact categories, reflecting a judgement about the relative importance of each impact category. For example, global warming and associated climate change may be perceived as a greater environmental threat than acidification. Such judgements are influenced by environmental concerns, but can also be influenced by natural, socioeconomic, or political factors for a particular nation or region.

The various steps in LCA are now discussed in greater detail, and illustrated by a case example.

## 10.4  Goal Definition

Firstly, it is important to define the purpose and context of the LCA exercise.

When comparing alternative processes or products, an appropriate basis for comparison of inventories and impacts must be selected, referred to as the *functional unit*. In comparing alternative processes for making a chemical product such as ammonia, a suitable functional unit would be 1 tonne of ammonia. In comparing processes to reduce $SO_2$ emissions to atmosphere, a tonne of sulphur dioxide removed from the effluent gas might be an appropriate basis.

When comparing alternative products, the task is more difficult. For insulating materials, insulating capability is a more important function than mass or volume of material. When comparing paint products, a suitable functional unit would need to incorporate the surface covered and the durability of the paint surface over time. A possible unit might be the volume of paint to cover and protect 1 $m^2$ of a particular surface over 10 years. Another example of a functional unit might be the packaging required for a given volume of beer. Note the volume of beer selected would determine whether the packaging was for a 'six-pack', a 'slab', or multiple slabs stacked on a pallet.

It is important to define the *organisational context* of the LCA study. For example, is the study to be made by a company for internal use to assess its competitive position, or for external use to shape customer perception? Is the study to be made on behalf of government for public use, for example, in assessing the impacts of alternative automobile fuels?

The *scope* of the LCA study encompasses the definition of the system under study including its boundaries, the level of detail and accuracy required, and the various data requirements.

System boundaries should be clearly defined and documented. Boundaries include both

- *space* considerations, for example, inclusive of all raw material mining and extraction, all transport operations, and

- *time* considerations encompassing product life and plant life, inclusive of decommissioning and waste storage time frameworks.

It is essential to include within the system boundary the impacts of utilities generation associated with the utilities consumed at the various stages of the product life cycle. Where utility generation is from a specific source, for example, steam supplied from a boiler at a chemical plant site, impacts can be identified and quantified. In other cases, for example, where electricity is drawn from a grid supply, impacts are usually derived from many sources of electricity supply and are accounted for on a weighted basis.

It may be important in some cases to estimate variability in input data, as well as an average value or a base case.

Modification of the study scope may be required as the LCA proceeds. A common constraint is difficulty in obtaining certain data. Simplifications may be justified where the accuracy of such data is not critical to the outcome.

Studies may be conducted for the whole life cycle, or restricted to certain parts of the life cycle. Thus, studies are often termed

- *cradle to grave*, encompassing the entire life cycle
- *cradle to gate*, encompassing raw materials extraction through a processing chain to a product leaving the factory gate (e.g., PVC manufacture inclusive of upstream manufacture of ethylene, chlorine, and vinyl chloride monomer, but exclusive of downstream fabrication, use, and disposal of PVC products)
- *gate to gate*, encompassing a manufacturing process at a particular site (e.g., manufacture of hydrogen gas from natural gas by the steam reforming process).

### 10.4.1 *Example of System Boundary Determination*

Figure 10.2 is a schematic representation of metal production from smelting a mineral sulphide ore with sulphur dioxide ($SO_2$) release and treatment. This is the context for a case study of alternative processes for desulphurising gases from metallurgical smelters (Golonka & Brennan, 1996). In this study, desulphurisation options included

- conversion of $SO_2$ to sulphuric acid
- reduction of $SO_2$ to sulphur, with option for subsequent sulphuric acid manufacture
- absorption of $SO_2$ in limestone slurry to form gypsum.

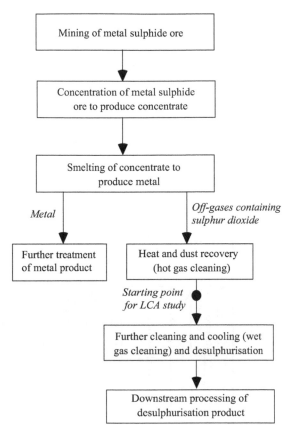

**Figure 10.2** Block diagram for smelting a mineral sulphide ore (Golonka & Brennan, 1996).

Options considered for downstream processing of sulphuric acid included

- manufacture of phosphoric acid
- reaction of phosphoric acid with ammonia to make ammonium phosphate fertiliser
- manufacture of superphosphate fertiliser from sulphuric acid and phosphate rock.

Figure 10.3 depicts the system boundary for an LCA study of the $SO_2$ treatment options. Activities within shaded area are included; activities outside shaded area are excluded. Some inclusions for the study were

- generation of energy, mainly electricity, steam and cooling water used in processes

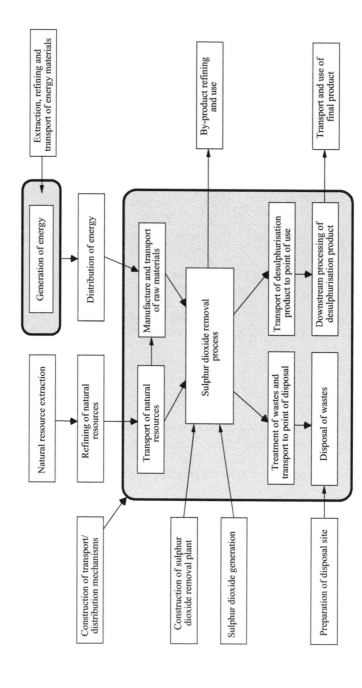

**Figure 10.3** System boundary definition for SO₂ treatment options (Golonka and Brennan, 1996).

- transport to the smelter of limestone, a raw material for one of the process options
- transport of desulphurisation products to markets
- disposal of wastes to the environment.

Some exclusions were

- mining of basic raw materials
- manufacture of catalysts
- extraction of natural gas from gas fields
- manufacture of materials for process plant and equipment.

A judgement was made that impacts associated with these excluded activities are relatively small, and that their assessment would add disproportionately to the time and cost of the study (i.e., much additional work would have been incurred for little gain in accuracy). It should be emphasised that there is always the opportunity to revise system boundaries to include one or more omitted impacts, as a study progresses.

Some effects were screened out from the assessment in the smelter gas study since they were common to all options, for example, removal of heavy metals in the wet gas cleaning step. Some key assumptions were applied to all processing options, for example, assumptions about the utility systems used.

## 10.5 Inventory Analysis

Inventory analysis involves the identification and quantification of all material and energy resources consumed and all wastes emitted, expressed in mass units. To achieve this, the process technology, design and operation of the plant must be considered and understood. One important data source is the *process flow-sheet*. The quality and detail of the flow-sheet and supporting work are important in determining the quality of the data. It is important to recognise that process flow-sheet analysis has limitations in that it excludes

- abnormal operation, start-up and shut-down procedures
- cleaning and maintenance activities
- fugitive emissions
- accidental releases, both minor and major
- losses from transport and related storage.

To overcome these limitations, plant and equipment design and plant operation must be considered, incorporating, for example,

- piping and instrumentation diagrams
- equipment specifications
- operating and maintenance procedures.

A separate source of inventory data to that from a process design perspective is performance data from an operating plant. It is important to recognise, however, that such data can vary from plant to plant due to variations in

- feedstock compositions
- differences in technology (often associated with age of the plant)
- process and plant design details
- operating and maintenance procedures.

Published inventory data is available for certain processes and products, for example, within commercial software packages for LCA and within published LCA studies, but the supporting basis for the data in such cases is often not fully provided. This can make data validation difficult.

### 10.5.1 *Treatment of Utilities and Energy*

In some published LCA studies, energy inputs are reported in energy units (e.g., MJ). Further definition requires quantification of resources consumed and wastes emitted through the use of fuels, electricity, and hot and cold utilities in processes. This further definition is needed as a study progresses, because impacts depend on the utility generation system used. For example, consumption of steam generated in a boiler from fuel combustion involves

- fuel consumption
- raw water consumption
- chemicals consumption in raw water and boiler water treatment
- waste water emissions from boiler blow-down
- flue gas emissions.

### 10.5.2 *Allocation Procedures*

In many systems or sub-systems, more than one useful product is produced. For example, in chlor-alkali plants, caustic soda and chlorine are the major products with smaller quantities of hydrogen. In ethylene plants, ethylene is the major product with lesser quantities of propylene and butadiene. In crude oil distillation within petroleum refineries, products include LPG, gasoline, aviation fuels, diesel, and fuel oil. How should inventories of resources and wastes and their associated environmental impacts be allocated between co-products in these cases?

The most common basis has been by relative mass flow of product, but economic value, volume, energy content, or molar flow have also been used. Since inventory data is in mass units, a mass basis has advantages from a fundamental scientific viewpoint. However, in some cases, the sales revenue from one product may greatly exceed that from a co-product, and one might argue that the more valuable product is driving the investment and should receive a greater allocation of the environmental burden. *The question of whether to use mass or economic value as the basis has caused considerable debate.*

## 10.6 Example of Inventory Data Estimation

Figure 10.4 is the result of flow-sheet analysis for production of sulphur or sulphuric acid from metallurgical smelter gases, and literature-based assessment for downstream processing of sulphuric acid to phosphate fertilisers (Golonka & Brennan, 1996). Sulphur can be converted to sulphuric acid, possibly at a separate location from the smelter. The reduced input of $SO_2$ to the direct production of $H_2SO_4$ shown on Fig. 10.4 enables all the data to be brought to a common basis downstream of $H_2SO_4$ production. The diagram quantifies

- raw materials and utilities consumptions
- products and wastes generated.

No allowance was made in this case study for fugitive or abnormal emissions. Fugitive emissions for $SO_2$ were judged likely to be negligible because of their toxic, pungent and readily detectable nature.

Assumptions were needed to assign inputs and outputs which correspond to consumptions of utilities. Assumptions common to all cases in the study were

- electricity was obtained from black coal
- fuel energy was obtained from fuel oil combustion
- water losses from evaporative cooling water circuits and steam circuits were 5% of the utility consumed.

Once inputs and outputs have been assigned to the utilities, inventory tables can be drawn up for the various processing steps.

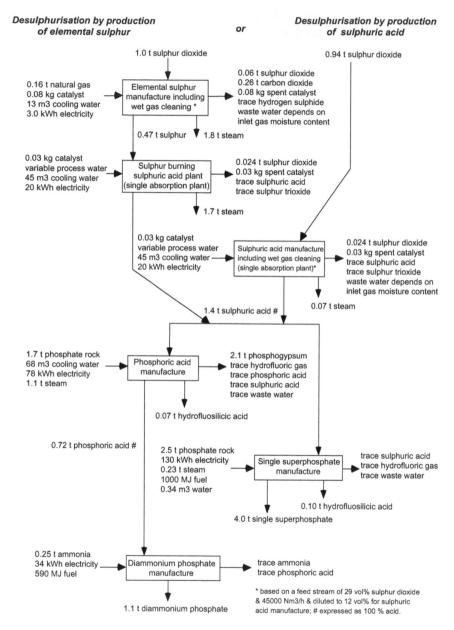

**Figure 10.4**   Inputs and outputs for SO$_2$ processing (Golonka & Brennan, 1996).

Table 10.1    Inventory data for conversion of $SO_2$ in smelter gas to sulphur (all values are in tonnes per tonne of sulphur dioxide treated) (Golonka and Brennan, 1996)

| | Inventory | Wastes (outputs) | Inventory |
|---|---|---|---|
| Major product | | | |
| Sulphur | 0.47 | | |
| Raw materials (inputs) | | | |
| Catalyst | 0.00008 | Ash | 0.0003 |
| Coal | 0.0021 | Carbon dioxide | −0.14 |
| Fuel oil | −0.13 | Nitrogen oxides ($NO_x$) | 0.00001 |
| Natural gas | 0.16 | Spent catalyst | 0.00008 |
| Water | 0.77 | Sulphur dioxide | 0.055 |

Table 10.1 summarises inventory data for the process where the $SO_2$ in smelter gas is converted to sulphur. In this process, $SO_2$ is reduced catalytically to sulphur with natural gas in two stages; the reactions are exothermic, making it possible for process heat recovery for steam generation. All inventory values are in tonnes per tonne of sulphur dioxide treated. Negative values for fuel oil consumption and carbon dioxide emissions result from credits claimed for by-product steam generated. It was assumed there was a demand for this by-product steam; if the steam required were not generated from process heat recovery, combustion of fuel would have been necessary. Fuel oil was chosen as a reference fuel in this study because of the remote sites typical for Australian mineral processing. Figure 10.5 is the result of mass and energy balances for generating 1 MJ of heat and 1 tonne of steam from fuel oil combustion. Assumptions for fuel oil included

- heat of combustion:          40.4 GJ/t fuel oil
- composition by mass of fuel oil:   85% carbon
                                2% sulphur
                                13% hydrogen.

## 10.7 Classification

Having established the inventory data, the next stage in LCA methodology is classification. The first step in classification involves allocating all inventory data for specific molecular species to particular impact categories. The terms used for these categories may differ slightly according to the particular form of LCA methodology used (see Table 10.2). Table 10.3 provides a

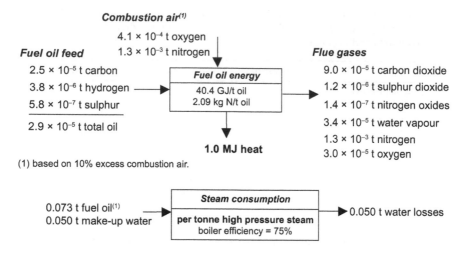

**Figure 10.5** Life cycle inventory for fuel oil combustion and for steam generation (Golonka, 1996).

basic description in an earlier UNEP manual of some more common impact categories. Victims (an additional category listed under CML in Table 10.2) relates to fatal consequences of an environmental disaster.

### 10.7.1 *Further Discussion of Impact Categories*

An expanded outline of the mechanisms and effects for more commonly encountered impact categories is now provided. More detailed accounts of these impacts, their causes and underlying mechanisms can be found in environmental science texts, for example, Masters (1998) and Masters and Ella (2008). Explanatory notes are also given in CML (1992, 2002).

**Global warming** has received much attention in the last 20 years and particularly in the last decade. Climate scientists are in agreement that there is a rise in the earth's temperature coincident with rising levels of carbon dioxide (and other greenhouse gases) in the atmosphere, attributable to anthropogenic causes. Carbon dioxide and the other greenhouse gases (predominantly methane, nitrous oxide and chloroflurocarbons (CFCs)) in the earth's atmosphere contribute to the 'greenhouse effect' by reducing the amount of heat radiated from the earth's surface. Likely effects of global warming on the earth include:

Table 10.2   Impact categories in life cycle assessment

| Heijungs (1992) (CML) | UNEP (1996) | SETAC (1993) |
|---|---|---|
| Enhancement of greenhouse effect | Global warming potential | Global warming potential |
| Acidification | Acidification potential | Acidification |
| Human toxicity | Human toxicity | Human toxicity |
| Photochemical oxidant formation | Photochemical oxidant potential | Photochemical oxidant formation |
| Depletion of biotic resources | | Depletion of biotic resources |
| Depletion of abiotic resources | Abiotic depletion | Depletion of abiotic resources |
| | Energy depletion | |
| Noise | | |
| Odour | | |
| Depletion of the ozone layer | Ozone depletion potential | Ozone depletion potential |
| Nutrification | Nutrification potential | Eutrophication |
| Ecotoxicity | Ecotoxicity — aquatic | Ecotoxicity |
| | Ecotoxicity — terrestrial | |
| Waste heat | | |
| Damage to ecosystems and landscapes | | Damage to ecosystems and landscapes |
| Victims | | |

- rising sea levels derived from the expansion of water in the oceans and from the melting of glaciers;
- increased intensity of the hydrological cycle with greater frequency and intensity of floods and droughts;
- deleterious effects on human health and the health of some ecosystems.

While a global phenomenon, these impacts of global warming are likely to be of different severity in different parts of the world. For example, low-lying islands will be vulnerable to rising sea levels, and communities near the delta regions of large rivers will be vulnerable to flooding.

**Photochemical oxidant potential**

Photochemical oxidant potential is related to photochemical smog, also referred to as urban smog, and historically as 'Los Angeles smog'. Photo-chemical smog is observed as brown air and is common in large cities with high traffic congestion and poor air ventilation. Photochemical smog is caused by the oxidation of volatile organic compounds and $NO_x$ in the presence of sunlight to form a range of pollutants with harmful effects on humans (particularly respiratory problems), vegetation, and a range of

Table 10.3   Brief explanation of impact category terms for some commonly encountered environmental impacts (UNEP, 1996)

---

**Abiotic, biotic,** and **energy depletion** are terms used under the umbrella of resource depletion. In early LCA work resource depletion was quantified by consumption divided by reserves.
**Abiotic depletion** — the extraction of non-renewable raw materials
**Biotic depletion** — the depletion of renewable resources. The rainforest and elephants are examples of biotic resources
**Energy depletion** — the extraction of non-renewable energy sources

**Global warming**
Increasing amounts of $CO_2$ and other greenhouse gases in the earth's atmosphere are leading to increased absorption of the radiation emitted by the earth and hence to global warming. $CO_2$, $N_2O$, $CH_4$, and CFCs contribute to global warming.

**Photochemical oxidant creation**
Under the influence of UV light, nitrogen oxides react with volatile organic compounds producing the photochemical oxidants that cause smog.

**Acidification**
Acid deposition, resulting from the release of nitrogen and sulphur oxides into the atmosphere, on soil and water can lead to changes in soil and water acidity, with effects on flora and fauna.

**Human toxicity**
Exposure to toxic substances — through air, water, or the soil, especially via the food chain — causes human health problems.
**Ecotoxicity** (aquatic and terrestrial) .Flora and fauna can be damaged by toxic substances, for example, cyanide. Ecotoxicity (in the UNEP system) is defined for both water (aquatic) and soil (terrestrial).

**Nutrification (or eutrophication)**
Addition of nutrients to soil or water increases the production of biomass. In water, this leads to reduced oxygen concentration which affects higher organisms such as fish. In soil and water, nutrification can lead to undesirable shifts in the number of species in ecosystems, thus threatening biodiversity.

**Ozone depletion**
Depletion of the ozone layer leads to an increase in the amount of UV-B radiation reaching the earth's surface. This radiation may result in some human diseases, damage to materials such as plastics, and interference with ecosystems.

---

materials. Pollutants formed include ozone, aldehydes, and peroxynitrates. The chemistry associated with the formation of photochemical smog is complex.

## Acidification

Release of acid gases into the atmosphere, primarily $SO_2$ and $NO_x$, lead to the formation of clouds of acidic water vapour which can drift beyond the sources of acid gas release. Precipitation of acid rain can then occur in nearby regions and countries leading to damage of waterways, forests, and buildings. Acid rain has been observed as a result of burning high sulphur coals, for example, in England with subsequent damage of forests in Switzerland and Germany, and in the United States with subsequent damage of forests in Canada. The pH of acid rain may be in the range 3–5 in contrast with that of unpolluted rain of 5–6 (mildly acidic because of $CO_2$ absorption).

## Nutrification

Nutrification (also referred to as eutrophication) is the process of nutrient enrichment in rivers, lakes, and waterways producing excessive vegetation, particularly algae. While a certain concentration of nutrients is desirable for supporting fish life, excessive amounts cause excessive amounts of algae, with a deterioration in water quality and destruction of fish life. Contributions to nutrification include industrial and domestic effluents as well as fertiliser carried by rain water run-off or groundwater seepage. The main chemical species contributing to nutrification are nitrogen and phosphorus.

## Ozone depletion

While ozone concentration in the troposphere (near ground level) is harmful and a major contributor to photochemical smog, ozone in the stratosphere (some 15 to 50 km above the earth's surface) has the beneficial effect of protecting the earth's surface from ultraviolet radiation. Increased exposure to ultraviolet radiation can potentially increase the risk of skin cancer in humans as well as causing damage to materials. CFCs and halons are inert and insoluble in water; thus, they have long atmospheric lifetimes, eventually drifting into the stratosphere where they cause destruction of ozone.

## Toxicity

Toxicity is assessed in terms of potential impacts to humans (human toxicity) and to aquatic, terrestrial, and sediment ecosystems (ecotoxicity).

Human toxicity assessment is based on human exposure to toxins derived from air, surface water, and soil. Different pathways are considered. For example, surface water contamination can affect human health through consumption of fish and drinking water. Soil contamination can affect human health through contamination of crops and cattle meat.

Models for the assessment of both human toxicity and ecotoxicity are complex and under continuous development and refinement. Human toxicity in the workplace has been extensively studied in the context of occupational health and safety.

**Resource depletion**

Depletion of resources such as minerals, fossil fuels, water, and biotic species such as fish, animals, birds, and plant life is difficult to quantify. Materials and energy consumption in industrial processes is a major contribution to resource depletion. Resource depletion is discussed in greater detail in Section 10.9.

## 10.7.2 *Assignment and Weighting of Chemical Compounds*

Some of the chemical compounds contributing to the inventory data are assigned to more than one impact category. For example, $SO_2$ is assigned to both acidification potential and human toxicity. $NO_x$'s are assigned to acidification, nutrification, and human toxicity. Within each impact category, *weightings* are assigned to the contributions of various compounds to impacts. For example, 1 tonne of $NO_x$ is reckoned to have 70% of the effect on acidification that 1 tonne of $SO_2$ has. Methane is reckoned to have 21 times the effect on global warming that carbon dioxide has. These relative weightings are also referred to as equivalency factors. Equivalency factors for selected compounds and selected impact categories are shown in Table 10.4. Embolded numbers indicate revisions made in the 2002 CML Handbook to data compiled in the 1992 CML Handbook. The extent of revisions reflects both uncertainty inherent in the weightings, and the ongoing research in environmental impact assessment.

Note that the number 1 is assigned to particular compounds in each impact category. Impact scores are then expressed in tonnes equivalent of these particular compounds. Thus, global warming potential is expressed in tonnes equivalent of $CO_2$, acidification potential in tonnes equivalent of $SO_2$, and so on. In some impact categories, for example, global warming potential, distinction is also made between impacts over different time horizons.

On the basis of inventory data and the equivalency factors, it is possible to develop an eco-profile enabling comparison of the impact potentials in the various environmental categories. Such eco-profiles can be drawn up for alternative processes, alternative stages of the product life cycle, or for the entire product life cycle.

It is important to distinguish the impact potentials thus calculated from the ultimate environmental impact at a particular site, which usually depends on site location and characteristics. For example, $SO_2$ emissions will have

Table 10.4   Equivalency factors for selected compounds, impacts (CML, 2002)

| Compound emissions | GWP 100 yr (kg/kg) | POP (kg/kg) | AP (kg/kg) | NP (kg/kg) | ODP (kg/kg) | HTair-100 yr (kg/kg) |
|---|---|---|---|---|---|---|
| Ammonia | | | 1.88 | 0.35 | | 1.0E-01 |
| Ammonium ion | | | | 0.33 | | |
| Benzene | | 0.218 | | | | 1.9E+03 |
| Butane | | 0.352 | | | | |
| Cadmium | | | | | | 1.5E+05 |
| **Carbon dioxide** | 1 | | | | | |
| **1,4 dichlorobenzene** | | | | | | 1 |
| Chromium VI | | | | | | 3.4E+06 |
| **CFC-11 refrigerant** | 4000 | | | | 1 | |
| CFC-12 refrigerant | 8500 | | | | 0.82 | |
| Ethane | | 0.123 | | | | |
| **Ethylene** | | 1 | | | | 6.4E-01 |
| Halon-1202 | | | | | 1.25 | |
| HCFC-22 refrigerant | 1700 | | | | 0.034 | |
| Hexane | | 0.482 | | | | |
| Hydrogen chloride | | | 0.88 | | | |
| Hydrogen fluoride | | | 1.6 | | | |
| Hydrogen sulphide | | | 1.88 | | | 2.2E-01 |
| Lead | | | | | | 4.7E+02 |
| Methane | 21 | | | | | |
| Mercury | | | | | | 6.0E+03 |
| Nitrate ion | | | | 0.1 | | |
| Nitric acid | | | 0.51 | 0.1 | | |
| Nitrogen | | | | 0.42 | | |

*(Contd.)*

Table 10.4    *(Contd.)*

| Compound emissions | GWP 100 yr (kg/kg) | POP (kg/kg) | AP (kg/kg) | NP (kg/kg) | ODP (kg/kg) | HTair-100 yr (kg/kg) |
|---|---|---|---|---|---|---|
| Nitrogen oxides NOx | | | 0.70 | 0.13 | | |
| Nitrous oxide N$_2$O | 310 | | | | | |
| Nitric oxide NO | | −0.427 | 1.07 | 0.2 | | |
| Nitrogen dioxide NO$_2$ | | 0.028 | 0.7 | 0.13 | | 1.2E+00 |
| Pentane | | 0.395 | | | | |
| **Phosphate** | | | | 1 | | |
| Phosphoric acid | | | 0.98 | 0.97 | | |
| Propane | | 0.176 | | | | |
| Styrene | | | | | | 4.7E-02 |
| **Sulphur dioxide** | 23,900 | | 1 | | | 9.6E-02 |
| Sulphur hexafluoride | | | | | | |
| Sulphur trioxide | | | 0.8 | | | |
| Sulphuric acid | | | 0.65 | | | |
| Trichloromethane | 4 | 0.023 | | | | 1.3E+01 |

**Note:** GWP denotes global warming, POP photochemical oxidant, AP acidification, NP nutrification and ODP ozone depletion potentials, while HT denotes human toxicity.

different impacts in forest, desert, or built-up city environments. Global warming, although a global impact, will have different effects on climate patterns and sea level impacts in different regions of the world. Much more data and analysis outside the traditional scope of LCA is needed to determine these impacts.

Not all impact categories will be significant in all studies. There may be energy intensive cases, for example, where impacts are dominated by global warming and resource depletion, and other effects might arguably be neglected and screened out early in the study. For the smelter gas desulphurisation study, four impact categories were judged to be dominant:

- acidification potential
- enhancement of greenhouse effect
- solid waste generation
- depletion of resources.

Table 10.5 summarises the classification step for sulphur production from smelter gas, using inventory data established earlier and listed in Table 10.1.

The simplest approach to calculating resource depletion (and that adopted by CML, 1992) is as the ratio of consumption to reserves. In the smelter gas treatment case, the consumptions of four resources are accounted for, including natural gas, coal, fuel oil, and water. Resource depletion for elemental sulphur production has been calculated as

$$\frac{\text{natural gas consumption from inventory table } (t/tSO_2)}{\text{world reserves of nature gas } (t)}$$

$$+\frac{\text{coal consumption from in inventory table } (t/tSO_2)}{\text{world reserves of coal } (t)}$$

$$+\frac{\text{oil consumption from inventory table } (t/tSO_2)}{\text{world reserves of water } (t)}$$

$$=\frac{0.16}{7.8 \times 10^{10}} + \frac{0.0021}{2.1 \times 10^{12}} + \frac{-0.13}{1.0 \times 10^{11}} + \frac{0.77}{4 \times 10^{13}}$$

$$= 1.02 \times 10^{-12}/tSO_2 \text{ treated}$$

The measure is only approximate. Estimating reserves is necessarily uncertain as it reflects what is discovered, what is economically recoverable, and the available technology for mining, extraction, and processing at the time of estimation.

Table 10.5　Classification step for sulphur production (Golonka and Brennan, 1996)

| Resource used | Catalyst | Natural gas | Coal | Fuel oil | Water | Total |
|---|---|---|---|---|---|---|
| Inventory, t/t $SO_2$ treated | 8.00E-05 | 0.16 | 0.0021 | −0.13 | 0.77 | |
| World reserves (tonnes) | Unknown | 7.80E+10 | 2.10E+12 | 1.20E+11 | 4.00E+13 | |
| Resource depletion, t/t $SO_2$ treated | Negligible | 2.05E-12 | 1.00E-15 | −1.08E-12 | 1.93E-14 | 1.0E-12 |

| Waste generated | Ash | $NO_x$ | $CO_2$ | $SO_2$ | Spent Catalyst | Total |
|---|---|---|---|---|---|---|
| Inventory, t/tSO₂ treated | 0.0003 | 1.00E-05 | −0.14 | 0.055 | 8.00E-05 | |
| Equivalency factor | | | | | | |
| Global Warming | | | 1 | | | |
| Acidification | | 0.7 | | 1 | | |
| Solid Waste | 1 | | | | 1 | |
| Impact, teq/t $SO_2$ | | | | | | |
| Global warming | | | −0.14 | | | −0.14 |
| Acidification | | 0.000007 | | 0.055 | | 0.055007 |
| Solid waste generation | 0.0003 | | | | 0.00008 | 0.00038 |

### 10.7.3 *Normalisation*

A further step following classification is the normalising of impacts in the various categories, according to regional or global contribution. For example, the acidification score for the process under consideration would be divided by the acidification score for that region, the global warming potential for the process by that for the region, and so on. This enables a comparison of the relative contributions of the process under study to the region's environmental profile.

It is possible to aggregate the various normalised impacts in different categories to provide an overall environmental index for the process under study. While useful, neither normalisation nor the aggregated index is an essential procedure in LCA methodology. The relative magnitudes of impact potential scores in separate categories for different regions reflect variations in natural, economic, and industrial systems. Thus, a normalised environmental index or eco-profile determined for a process can differ from one region or country to another.

Table 10.6 shows the normalised impacts for sulphur production in the $SO_2$ case calculated using Australian consumption and emission data available at the time of the study. Normalised impacts are expressed in units of yr/t $SO_2$ (i.e., reciprocal of t $SO_2$/yr). The normalised impact scores for this study identified acidification and resource depletion as the two major impact categories for converting $SO_2$ to sulphur. Normalised impacts may be further multiplied by the annual $SO_2$ emission from a particular smelter source to reflect the scale of the impact potential. In this latter case, the normalised impacts are dimensionless. Table 10.7 shows impact category scores, normalised and aggregated for all desulphurisation options, normalized and scaled for annual emissions from a smelter of 244 ktonnes $SO_2$/yr.

## 10.8 Improvement Analysis

By comparing indices or inventory data for processes or products, it is possible to identify those phases of product manufacture or use which make major contributions to environmental impact. It is also possible to identify what impact category contributes most to environmental damage, what molecular species it derives from, and where in the product life cycle the inventory is created. Product design, process design, or operations strategy can then be reviewed and modified to reduce impacts.

Table 10.6 Normalised impacts for sulphur production using Australian consumption and emission data (Golonka and Brennan, 1996)

| Impact category | Impact score for process per t $SO_2$ treated | Australian resource depletion | Australian emission /yr | Normalised impact, yr/t $SO_2$ treated |
|---|---|---|---|---|
| Resource depletion | 9.88E-13 | 5.24E-04 | | 1.88E-09 |
| Global warming | −1.40E-01 t $CO_2$eq | | 5.76E+08 t eq $CO_2$ | −2.43E-10 |
| Acidification | 5.50E-02 t $SO_2$eq | | 2.23E+06 t eq $SO_2$ | 2.47E-08 |
| Solid waste generation | 3.84E-04 t | | 6.60E+08t | 5.76E-13 |
| Total | | | | **2.63E-08** |

Table 10.7 Impact category scores for all alternative desulphurisation processes, normalised and scaled (Golonka and Brennan, 1996)

| LCA category | Australian impact (2) | Normalised impact score for process option (1) | | | |
|---|---|---|---|---|---|
| | | Sulphuric acid | Elemental sulphur | Saleable gypsum | No $SO_2$ removal |
| Acidification potential | $2.2 \times 10^6$ (t/yr) | $2.7 \times 10^{-3}$ | $6.0 \times 10^{-3}$ | $3.5 \times 10^{-3}$ | $1.1 \times 10^{-1}$ |
| Global warming | $5.8 \times 10^8$ (t/yr) | $6.2 \times 10^{-6}$ | $-6.1 \times 10^{-5}$ | $3.9 \times 10^{-4}$ | 0 |
| Solid waste generation | $6.6 \times 10^8$ (t/yr) | $7.7 \times 10^{-7}$ | $1.4 \times 10^{-7}$ | $6.8 \times 10^{-5}$ | 0 |
| Depletion of resources | $5.4 \times 10^{-4}$ (/yr) | $1.5 \times 10^{-5}$ | $4.5 \times 10^{-4}$ | $6.1 \times 10^{-5}$ | 0 |
| Total index | | $2.8 \times 10^{-3}$ | $6.4 \times 10^{-3}$ | $4.0 \times 10^{-3}$ | $1.1 \times 10^{-1}$ |

(1) Calculated as a ratio of $\dfrac{\text{value of category for processing option}}{\text{value of category for Australia}} \times 244$ kt/yr $SO_2$.

(2) Calculated from total Australian emissions and consumptions.

For the smelter case study, comparison of the total indices leads us to the following conclusions:

1. All treatment options are preferred to the no treatment option.
2. For the desulphurisation options, sulphuric acid is preferred ahead of gypsum ahead of elemental sulphur.

However, there are still questions to be answered.

- Sulphuric acid is an intermediate and must be further processed. What are the environmental impacts then incurred?
- What are the environmental impacts incurred in the use of the resulting downstream products?
- What are the market and cost factors associate with transportation, further processing and product use?

Some of these issues are discussed in Golonka and Brennan (1996, 1997) for the $SO_2$ treatment study, and in a more general sense in later chapters of this book.

## 10.9 Some Challenges and Uncertainties in LCA

We now review some challenges and uncertainties faced when applying LCA methodology to different case studies.

### 10.9.1 *Goal Definition*

Boundary definition is a key aspect in all LCA studies. The decision may be influenced by perceived relative importance of effects, as well as by availability of data. Boundary definition can be an iterative decision during a study, with refinements made as increased information or better understanding are obtained. Unfortunately, for many published LCAs there is lack of clear definition of the system boundary used.

### 10.9.2 *Inventory Data*

There are a number of obstacles to obtaining inventory data for a process or product. These include for example

- the proprietary nature coupled with limited understanding of many process technologies

Table 10.8    Fuel impacts in road and rail transport (Bureau of Transport and Communication Economics, 1995)

| Impact | Road | Rail |
|---|---|---|
| Diesel consumption | 1.2 MJ/t.km | 0.13 MJ/t.km |
| Carbon dioxide emissions | 70 g/MJ fuel | 70 g/MJ fuel |
| Carbon monoxide emissions | 7.9 g/km | 0.58 g/MJ fuel |
| Hydrocarbon emissions | 2.8 g/km | 0.13 g/MJ fuel |
| Nitrogen oxide emissions | 15 g/km | 1.7 g/MJfuel |
| Particulate emissions | 0.8 g/km | 1.5 g/km |
| Sulphur dioxide emissions | 1.4 g/km | 0.083 g/MJ fuel |

- relatively recent development of LCA methodology coupled with limited availability of suitable personnel to develop the data.

Where published data exists, there is often a lack of transparency regarding its source, basis, and accuracy.

Inventory data does not account for concentration effects, either in an effluent or in the receiving medium. Such concentrations can influence the environmental damage arising from emissions.

A developed process flow-sheet accounts only for design conditions and steady-state case. More detailed studies supported by operating experience is necessary to account for

- fugitive emissions
- accidental spills or releases
- emissions under start-up, shut-down, abnormal operating, and maintenance conditions.

Inventory data makes no distinction between mobile (e.g., transport) emissions and stationary emissions. Transport of raw materials or products in most cases involves fuel combustion. While the emissions associated with transportation will be distributed over a large area, emissions from stationary sources are concentrated in one location. Some reported data for fuel related impacts in materials transport is listed in Table 10.8.

Road data listed here is for 'non-urban articulated trucks' and rail data is for private freight, both using automotive diesel oil. In order to use the truck data presented in the $SO_2$ treatment study, a load of 42.5 tonnes was assumed per truck trip, consistent with assumptions made in the quoted report.

Data for *shipping* (Bureau of Transport and Communication Economics, 1992) suggests that for a consumption of 3.0 tonnes of marine diesel fuel per 1000 DWT, the following emissions are approximated:

| | |
|---|---|
| Carbon dioxide emissions | 27 g/tonne km |
| Carbon monoxide emissions | 8 kg/tonne fuel |
| Hydrocarbon emissions | 2.7 kg/ tonne fuel |
| Nitrogen oxide emissions | 59 kg/tonne fuel |
| Sulphur dioxide emissions | 21 × S kg/tonne fuel where S is % wt of sulphur in fuel. |

### 10.9.3  By-Products — Marketable or Waste?

An important issue is the identification of output streams as product or waste. This decision is heavily geared to market criteria, and emphasises the role of market analysis as a subset of environmental assessment. In the case of hydrocarbon processing, co-products can alternatively be sold or burned, with or without energy recovery, depending on available markets. Ethylene manufacture from naphtha is a specific example, where co-products include propylene, butadiene, and pyrolysis gasoline. Burning involves the liberation of combustion products. However, if the related energy can be recovered, there are benefits through the avoided impacts of resource depletion and wastes derived from alternative fuel use.

Solid wastes often offer greater potential for future marketability than gaseous or liquid wastes because of their greater ease of storage and transport.

### 10.9.4  Impact Analysis

In the application of LCA methodology, a distinction is made between impact potentials often referred to as 'midpoint categories', and the subsequent damage to human health or the environment often referred to as 'endpoint categories'. Endpoint damage reflects the influence of *location* in determining specific impacts. Examples of interest in the $SO_2$ treatment case include

- access to a disposal site for disposal of non-hazardous waste, for example, disposal of gypsum at the remote mine site or in the city
- effect of site on environmental effects of an emission, for example, effect of $SO_2$ on desert, forest, or built-up city environments.

Midpoint categories are useful in process comparisons made in isolation from site location considerations. However, they are of less value once site location options for a project are being compared.

### 10.9.5 *Resource Depletion*

*Materials and energy consumptions* are of key importance both in terms of resource depletion, and the environmental impacts associated with resource extraction, refining, and transport prior to use. Some specific burdens associated with resource use can be identified:

- the burden associated with exploration for a resource and proving its viability
- the burden associated with extraction, refining and transport of a resource
- the burden of site rehabilitation after certain mining and extraction activities
- the consequences of loss of the consumed resource
- the burden associated with replacement of the consumed resource with alternatives.

An early impact measure adopted for abiotic resource depletion was consumption divided by reserves (Heijungs, 1992). The concept of reserves is blurred, however. There is also a distinction between *ultimate* and *economic* reserves. *Ultimate reserves* comprise the total quantity of reserves, not only in ores, but also present in the earth's crust. For example, aluminium is present in 8% of the earth's crust, largely as aluminium silicates. However, aluminium is mainly extracted from bauxite ore, present as a much smaller percentage of the earth's crust, but as a much richer source of alumina. *Economic reserves* comprise that component of a resource which can be economically recovered with the technology and cost/price structure of the day.

It is difficult to account for a decline in quality of reserves over time and for cost and environmental impacts of recovering resources in the future. For example, there has been a sustained increase over time in the density and sulphur content of crude oil, with implications for costs and environmental impacts in refining. As the mineral content of metal bearing deposits decreases, increased energy is required for mining, crushing, milling, and beneficiation per tonne of metal, with implications for greenhouse gas emissions.

It is also difficult to predict the success of future exploration activities, and the possibility that future deposits of fossil fuels or minerals might be discovered in environmentally sensitive areas, with attendant risks and costs.

In assessing resource depletion, there is a question of the ease of substitution for different resources, for example, coal, uranium, oil, and gas in the energy resource sector, and for different end product applications, such as construction materials and packaging materials.

One definition of resources makes distinctions between three categories (de Haes *et al.*, 1999):

- deposits which cannot be replaced (e.g., fossil fuels and mineral ores)
- funds, which are replaced naturally but over time (e.g., groundwater, sand and clay)
- flows, which are continuously and completely replaced (e.g., solar energy, wind and surface water).

Some flexibility has been introduced (Guinee, 2002) for impact assessment of abiotic resource depletion. Ultimate reserves, economic reserves, or exergy content of resource are some of the alternative measures suggested.

Supplementary indicators for resource depletion might be useful in specific case studies (Brennan, 2005). These include location, accessibility, quality, potential for substitution, alternative uses, market value, and opportunity cost. Table 10.9 shows suggested indicators for Australian fossil fuels used in power generation.

## 10.9.6 *Normalisation*

Regional and global impact data used for normalisation is necessarily limited in its reliability. Some earlier examples of global data were merely extrapolations of more reliable data for specific countries (e.g., Holland) or regions (e.g., Europe), based on relative populations. Some examples of data sources for normalisation used in various studies are listed in Table 10.10.

In the sulphur dioxide study by Golonka and Brennan (2006), normalisation was performed using data for Australia, the United States, Canada, and Sweden. Relative magnitudes of normalised impacts were found to be insensitive to these regions. However, in a petroleum desulphurisation study by Burgess and Brennan (2001), discussed in Chapter 11, the normalised environmental profile was sensitive to the region considered. For Holland, Switzerland, or Sweden, the normalised impact score for photochemical oxidant potential was dominant in the eco-profile. However, for Australia, normalised impact scores for global warming and energy depletion exceeded that for photochemical oxidant potential.

Table 10.9    Resource indicators for Australian fossil fuels in power generation

| Resource | Brown coal | Black coal | Natural gas |
|---|---|---|---|
| Location | Latrobe Valley, Vic | Qld, NSW, WA | Bass Strait, NW Aust, Cooper Basin |
| Accessibility | Useful only at source | Transportable to power station by rail or ship | Transportable to power station by gas pipeline or LNG shipping[1] |
| Quality issues | $H_2O$, S, N, Ash, other, low H to C ratio | S, N, Ash, other Low H to C ratio | $CO_2$, $H_2S$, $H_2O$, Hg High H to C ratio |
| Environmental issues[2] | High $CO_2$ emissions per MWh | High emissions of $CO_2$ per MWh | Relatively low $CO_2$ emissions per MWh |
| Potential for Substitution | Gas, black coal, crude oil. Limited potential for renewables | Gas, brown coal, crude oil. Limited potential for renewables | Coal, crude oil Limited potential for renewables |
| Alternative uses | Limited with present technology | Limited with present technology | Fuel for heating, drying. Feedstock for chemicals (e.g., $H_2$, $CH_3OH$, HCN) |
| Market value | Very low | $50–100/tonne | $2–5/GJ |
| Opportunity cost[3] | Close to zero | Medium | High |

[1] Long distances from NW Shelf to east coast markets make transport costly.
[2] Life cycle assessments for range of technologies detailed in Table 11.1 in Chapter 11.
[3] High moisture content of brown coal makes transport economically non-viable.

### 10.9.7 *Valuation*

The valuation phase requires weighting impact category scores in terms of their relative environmental importance. However, the relative importance of damage categories is difficult to assess on an objective scientific basis. Public, political, and scientific opinion may be canvassed as a basis for allocating weightings. However, the primary focus of attention can shift as evidence of different types of environmental damage are progressively discovered and addressed. Thus,

- acidification effects were recognised in forests in the late 1970s in Norway, Germany, and the United States leading to increased regulation and funding for mitigation

Table 10.10  Selected normalisation data for various countries and regions

| Region reference | | World | The Netherlands | The Netherlands | Western Europe | Australia | Australia | Australia | Australia |
|---|---|---|---|---|---|---|---|---|---|
| Impact category | Unit | 1 | 1 | 2 | 2 | 3 | 4 | 5 | 6 |
| Natural gas consumed | kg/yr | | | | | **9.20E+11** | | | 2.36E+13 |
| Water consumption | kg/yr | | | | | **3.52E+14** | | 1.46E+13 | |
| Lignite consumption | kg/yr | | | | | 4.31E+13 | | | 5.30E+13 |
| Black coal consumption | kg/yr | | | | | **1.91E+11** | | | 2.52E+14 |
| Energy depletion | GJ/yr | 2.35E+11 | 2.35E+09 | | | | | 4.20E+09 | |
| Global warming | kg/yr | 3.77E+13 | 3.77E+11 | 2.10E+11 | 4.20E+12 | 4.31E+11 | 5.76E+11 | 4.03E+11 | 4.78E+11 |
| Photochemical oxidant | kg/yr | 3.74E+09 | 3.74E+07 | 1.90E+09 | 6.30E+09 | 1.56E+09 | | 8.95E+08 | 9.88E+08 |
| Acidification | kg/yr | 2.86E+11 | 2.86E+09 | 9.20E+08 | 3.40E+10 | 3.36E+09 | 2.23E+09 | 2.94E+09 | 3.47E+09 |
| Human toxicity | kg/yr | 5.76E+11 | 5.76E+09 | 8.80E+08 | 3.90E+10 | | | | |
| Aquatic ecotoxicity | m3/yr | 9.08E+14 | 9.08E+12 | 8.90E+12 | 4.40E+14 | | | | |
| Terrestrial ecotoxicity | kg/yr | 1.16E+15 | 1.16E+13 | 1.20E+13 | 2.30E+14 | | | | |
| Nutrification | kg/yr | 7.48E+10 | 7.48E+08 | 1.10E+09 | 8.60E+09 | | | | 3.44E+08 |
| Ozone depletion | kg/yr | 1.00E+09 | 1.00E+07 | 4.40E+06 | 5.60E+07 | | | | |
| Solid waste disposal | kg/yr | | | 8.80E+09 | | **6.06E+10** | 6.60E+11 | | 2.12E+13 |

**References**

1. UNEP (1996)
2. Blonk (2005)
3. Williams (2002)
4. Golonka and Brennan (1996)
5. Burgess (1999)
6. May (2003)

- ozone layer depletion resulted in the signing of the Montreal Protocol (1988)
- global warming, while generally regarded as less important in the 1970s and 1980s, became intensely debated in the late 1990s, for example, at the Tokyo summit, and continues to be debated.

## 10.10  Some Alternative or Supplementary Approaches to LCA

### 10.10.1  *EPS System — An Example of Evaluation Used with Inventory Data*

The Environmental Priority Strategies (EPS) system was developed jointly by the Swedish Environmental Institute and Volvo. The EPS system assigns a multiplier to each element of inventory data to determine an environmental load. For example, if production of 1 kg ethylene results in emission of 0.53 kg $CO_2$ and 0.006 kg $NO_x$, the environmental load units (ELUs) for these two inventory entries are

$$0.53 \times \mathbf{0.09} = 0.048 \quad \text{ELU for } CO_2 \text{ and}$$
$$0.006 \times \mathbf{0.22} = 0.0013 \quad \text{ELU for } NO_x$$

where 0.09 and 0.22 are the multiplier indices recommended by the EPS system for $CO_2$ and $NO_x$, respectively. The EPS referred to these multipliers as 'environmental indexes'. By summing the ELUs over all inventory elements, a total ELU for a product can be estimated.

A selection of multipliers for EPS 1996 are published in Allen and Rosselot (1997). Volvo has used the EPS system to evaluate product designs and alternative materials.

EPS 2000 is an update of EPS 1996. Five areas of environmental protection are considered:

- human health
- ecosystem production capacity
- abiotic stock resources
- biodiversity
- cultural and recreational values.

Weighting factors, based on the contingent valuation method of willingness to pay to preserve environmental quality, are used.

Note that the EPS approach represents a single step between the Inventory and Evaluation stages of an LCA study. The EPS system inevitably reflects a location bias and some subjective assessment.

### 10.10.2 *Eco-Indicator*

Another approach is the Eco-Indicator. Eco-Indicator 99 links inventory data through damage potentials to endpoint impact categories of

- damage to human health
- damage to ecosystem quality
- damage to fossil and fuel resources.

The Eco-Indicator and EPS approaches have been discussed and reviewed by Azapagic (2006). Further information is also available for each approach from http://www.pre.nl/simapro/impact_assessment_methods.htm.

## 10.11  LCA Software

A number of commercial software packages are available for LCA studies. Some of these have inbuilt databases. The transparency of system boundaries and key assumptions for data are important considerations when using these packages.

Some better known packages include GaBi developed by PE International www.pe-international.com and Sima Pro developed by Pre consultants http://www.pre.nl/.

## 10.12  Concluding Remarks

While there are limitations and uncertainties in LCA methodology, its structured approach, scientific basis, and quantitative nature make LCA a valuable tool for the environmental assessment of products and processes. Valuable features include the life cycle approach, and the wide spectrum of impacts derived from consumption and emissions. Inventory analysis derives directly from process system definition and from mass and energy balances, where chemical engineering can make a unique contribution.

LCA methodology has undergone considerable development in the last 20 years and there has been a growing acceptance of its use by industry, government and the professions. Case study work has an important role in understanding the basis and application of LCA. There is opportunity to link other approaches with LCA for particular studies, for example, economic and sustainability assessments, and site-specific environmental impact assessments.

As the collective experience in applications grows, as more data becomes available, and scientific understanding of environmental impacts grows, LCA will become an increasingly valuable tool for environmental assessment.

## Acknowledgements

The case study on sulphur dioxide emission abatement is drawn from the work of K. A. Golonka, his thesis and joint publication with D. J. Brennan.

Figures 10.2, 10.3, 10.4, and Tables 10.1, 10.5, and 10.7 have been reprinted from Golonka K. A. and Brennan D. J., *Application of Life Cycle Assessment to Process Selection for Pollutant Treatment: A Case Study of Sulphur Dioxide Emissions from Australian Metallurgical Smelters* in Transactions of the Institution of Chemical Engineers, Volume 74, Part B, May 1996, 105–119, with permission from Institution of Chemical Engineers.

## References

Allen, D. T. and Rosselot, K. S. (1997) *Pollution prevention for Chemical Processes*, John Wiley and Sons, New York.

Azapagic, A. (2006) Life cycle assessment as an environmental sustainability tool, in *Renewables-Based Technology* (ed. Dewulf, J., and Van Langenhove, H.), John Wiley and Sons, Chichester, England.

Blonk, H. (2005) Three reference levels for normalization in LCA at www.pre.nl.

Brennan, D. J. (February 2005) *Sustainability Measures for Decision Support: Four Outstanding Problems for Resolution*, Proceedings of 4th Australian Life Cycle Assessment Conference, Sydney.

Bureau of Transport and Communication Economics (1992) *Fuel Efficiency of Ships and Aircraft*, Working paper 4, Australian Government Publishing Service, Canberra.

Bureau of Transport and Communication Economics (1995) *Greenhouse Gas Emissions from Australian Transport : Long Term Projections*, Report 88, Australian Government Publishing Service, Canberra.

Burgess, A. (1999) *An Environmental and Cost Assessment of Gas Oil Desulphurisation*, MEngSc thesis, Monash University.

Burgess, A., and Brennan, D. J. (2001) Environmental and economic analysis of petroleum desulphurisation, *J Clean Prod*, **9**(5), 465–472.

Burgess, A. and Brennan D. J. (2001) Application of life cycle assessment to chemical processes, *Chem Engg Sci*, **56**(8), 2589–2604.

de Haes, U. H. A, Jolliet, O., Finnveden, D., Hauschild, M., Krewitt, W., and Muller-Wenk, R. (1999) Best available practice regarding impact categories and category indicators in life cycle assessment, *Int J Life Cycle Assess*, **4**(2), 66–74 and (3), 167–174.

Golonka K. A. and Brennan D. J. (1996) Application of life cycle assessment to process selection for pollutant treatment: a case study of sulphur dioxide emissions from Australian metallurgical smelters, *Trans Inst Chem Eng*, **74**(Part B), 105–119.

Guinee, J. B. (2002) *Handbook on Life Cycle Assessment*, Centre of Environmental Science, Leiden, Kluwer Academic Publishers, Dordrecht, the Netherlands.

Heijungs, R. ed. (1992) *Environmental Life Cycle Assessment of Products*, Centre of Environmental Science, Leiden, the Netherlands.

Masters, G. M. (1998) *Introduction to Environmental Engineering and Science*, 2nd edn, Prentice Hall, Upper Saddle River, New Jersey.

Masters, G. M. and Ela, W. P. (2008) *Introduction to Environmental Engineering and Science*, 3rd edn, Pearson Education, Upper Saddle River, New Jersey.

May, J. (2003) *Sustainability of Electricity Generation using Australian Fossil Fuels*, PhD thesis, Monash University.

SETAC (1993) *Guidelines for LCA: A 'Code of Practice'*, Leiden, the Netherlands.

UNEP (1996) *LCA: What it is and How to do it*, United Nations, Paris.

Williams, S. (2002) *An Application of LCA to an Australian Ethylene Production Facility*, M.Eng.Sc thesis, Monash University.

# Chapter 11

# Life Cycle Assessment Case Studies

## 11.1 Introduction

In Chapter 10, we explored the methodology of life cycle assessment (LCA) in conjunction with a case example on the treatment of sulphur dioxide emissions from metallurgical smelters. In this chapter, we explore some additional case studies drawn from previous work dealing with life cycle inventory data for utilities and an LCA study of diesel desulphurisation.

## 11.2 Life Cycle Inventories for Common Utilities

As part of LCA studies, inventory data are needed for utilities in terms of resources consumed and wastes emitted. For many products and processes impacts derived from utilities dominate the environmental impact. The following examples, drawn from work by Golonka (1996) and Burgess (1999) show the assumptions made and the derived utility data. *Impacts in mining, extraction, and transport of fuel to the utility generation plant have been excluded* due to their variability. The inventory data are not intended to reflect the situation for all cases; fuel and water qualities and utility systems may also vary considerably from case to case.

*Sustainable Process Engineering: Concepts, Strategies, Evaluation, and Implementation*
David Brennan
Copyright © 2013 Pan Stanford Publishing Pte. Ltd.
ISBN 978-981-4316-78-1 (Hardcover), 978-981-4364-22-5 (eBook)
www.panstanford.com

## 11.2.1 *Assumptions by Golonka and Burgess Regarding Utility Systems*

**Electricity generation (Golonka, 1996)**

- electricity is obtained from Australian black coal
- black coal contains 16.7 wt % ash, 0.5 wt % sulphur, 60 wt % fixed carbon; fuel value is 24 GJ per tonne of coal; 2.36 kg nitrogen is emitted per tonne coal; 90% of the nitrogen is emitted as NO, 10% as $NO_2$
- efficiency of power generation is 36%.

**Cooling water generation (Golonka, 1996)**

- 5% of circulation rate is required as make-up water for cooling water circuit
- electric power of 0.182 $kWh/m^3$ is consumed by air fans (estimated from Peters & Timmerhaus, 1991) and circulating pumps
- cooling water temperature is 28°C, leaving the cooling towers with a 20°C temperature rise in heat exchangers.

**Steam generation (Burgess, 1999)**

- steam is generated using natural gas as fuel
- 5% of steam produced is required as water make-up for steam circuit
- boiler efficiency, defined as the ratio of energy used in generating steam to energy liberated from fuel combustion, is 75%.

*For low pressure steam*, conditions of steam are 133°C, 300 kPa, *h* (saturated steam) = 2725 kJ/kg. *For high pressure steam,* conditions of steam are 250°C, 40 bar, *h* (saturated steam) = 2801 kJ/kg.

**Natural gas (Burgess, 1999)**

- lower calorific value of natural gas is 46.7 MJ/kg
- natural gas composition: 90.6 % vol $CH_4$, 5.6 % vol $C_2H_6$, 0.8 % vol $C_3H_8$, 0.2 % vol $C_4H_{10}$, 1.1 % vol $N_2$, 1.7 % vol $CO_2$, trace $H_2S$ (45 mg $S/Nm^3$ max)
- all sulphur in feed is assumed converted to $SO_2$
- $NO_x$ emission has been assumed as 0.06 kg/GJ for natural gas fired heaters of up to 30 MW duty, and as 0.08 kg/GJ of fuel energy combusted in gas fired steam generation (Bently & Jelinek, 1989).

## 11.2.2 *Derived Inventory Data*

Derived inventory data are summarised in Figs. 11.1 to 11.3.

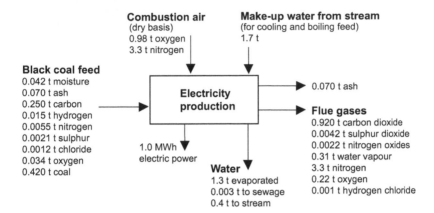

**Figure 11.1** Inventory data for electricity from black coal. Data obtained from Golonka (1996).

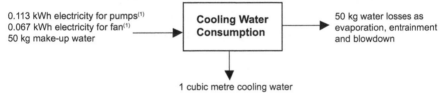

**Figure 11.2** Recirculated cooling water incorporating cooling tower. Data obtained from Golonka (1996).

## 11.3 Inventory Data for Distinct Electricity Supply Systems

In subsequent work by May (2004), inventory data were derived for electricity generation from a range of fossil fuels and a range of electricity generation cycles. *In this work, system boundary incorporated fuel extraction and purification, fuel transport to the power station, electricity generation and electricity transmission to the point of use.* Power generation technologies included those currently in common use and some promising advanced technologies for coal, specifically integrated gasification combined cycle. The cases of exporting Australian LNG and black coal to Japan for power generation in that country were also considered.

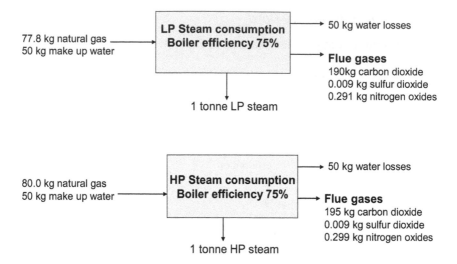

**Figure 11.3** Inventory data for steam generation from natural gas. Data obtained from Burgess (1999).

Inventory data encompassing key inputs and outputs for 12 fuel/technology combinations were developed and used to estimate normalised environmental impact indices, applying LCA methodology.

Table 11.1   Input and output inventory data for extended electricity supply systems; units are in kg per MWh delivered (adapted from May and Brennan, 2002)

| System | ST-BrC | ST-BlC | OCGT-NG | IGCC-BrC | IGCC-BlC | CCGT-NG | CCGT-LNG |
|---|---|---|---|---|---|---|---|
| *Inputs* | | | | | | | |
| Main fuel | 1464 | 467 | 303 | 754 | 261 | 159 | 178 |
| Limestone | | | | | 35.7 | | |
| *Outputs* | | | | | | | |
| Solids | | | | | | | |
| Ash | 21.8 | 87.3 | | 6.8 | 26.7 | | |
| Gases | | | | | | | |
| $CO_2$ | 1209 | 1023 | 825 | 740 | 660 | 445 | 535 |
| $CH_4$ | 0.007 | 1.65 | 0.979 | 0.340 | 0.974 | 0.478 | 1.28 |
| NMVOC | 0.022 | 0.024 | 0.048 | 0.011 | 0.012 | 0.174 | 0.475 |
| $N_2O$ | 0.018 | 0.009 | 0.006 | 0.011 | 0.012 | 0.010 | 0.014 |
| $NO_x$ | 1.68 | 3.36 | 2.83 | 0.187 | 0.200 | 0.767 | 1.29 |
| CO | 0.223 | 0.142 | 0.685 | 0.211 | 0.167 | 0.096 | 0.228 |
| $SO_2$ | 1.91 | 4.17 | 0.045 | 1.93 | 0.059 | 0.026 | 0.100 |
| Particulates | 0.493 | 0.548 | 0.285 | 0.022 | 0.038 | 0.037 | 0.063 |

**Note:** Flue gas scrubbing for $SO_2$ removal has been assumed for IGCC-BlC

Table 11.1 shows selected inventory data from these studies. Fuel and generation system nomenclature adopted for Table 11.1 is as follows:

| | |
|---|---|
| ST-BrC | Steam turbine using brown coal |
| ST-BlC | Steam turbine using black coal |
| OCGT-NG | Open cycle gas turbine using natural gas |
| IGCC-BrC | Integrated gasification combined cycle using brown coal |
| IGCC-BlC | Integrated gasification combined cycle using black coal |
| CCGT-NG | Combined cycle gas turbine using natural gas |
| CCGT-LNG | Combined cycle gas turbine using liquefied natural gas |

## 11.4 Hydrotreating of Diesel

### 11.4.1 *Hydrotreating Process*

Hydrotreating is the key operation in petroleum refining for control of sulphur contents of fuels. Tighter targets for sulphur in gasoline and diesel have been adopted worldwide in recent years. This study examines the hydrotreating of gas oil to produce diesel blend stock. The primary aim was to identify environmental burdens incurred, their relative importance, and their source. Reactor temperature and pressure for the hydrotreater are of the order of 350°C and 60 bar, though these conditions vary depending on feedstock composition and the catalyst used. In the reactor, sulphur in various forms within the hydrocarbon feedstock is converted by hydrogen to hydrogen sulphide. Hydrogen compressors and hydrocarbon feed pumps are used to achieve reactor pressure. A fired heater is used to heat the combined feed to achieve reactor temperature. After the reactor, heat is recovered and pressure is let down to enable separation of unreacted hydrogen from hydrocarbons. Unreacted hydrogen is recycled while the hydrocarbons are distilled to separate light hydrocarbons and hydrogen sulphide ($H_2S$) from the desulphurised product. The light hydrocarbons are sent to an amine absorption/stripping plant to recover $H_2S$, for subsequent conversion to sulphur in a sulphur recovery unit.

### 11.4.2 *System Boundary*

The system boundary adopted for the study and the linkages between processing units are shown in Fig. 11.4. The system includes the following process plants:

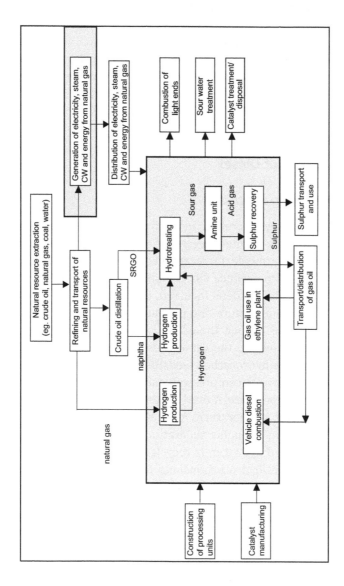

**Figure 11.4** System boundary diagram for hydrotreating study (Burgess and Brennan 2001).

- hydrotreating unit where gas oil is treated with hydrogen at elevated temperature and pressure in presence of a catalyst, converting sulphur to $H_2S$
- hydrogen production plant
- amine absorption and stripping plant to recover $H_2S$
- sulphur recovery units where $H_2S$ is converted to sulphur.

Two alternative sources of hydrogen were considered in the study:

- steam reforming of natural gas in a dedicated hydrogen plant (SRU)
- hydrogen by-product from a catalytic reformer unit within the refinery (CRU).

Shaded areas on the diagram indicate operations included in the LCA study. Since the product in this case was used both for diesel fuel and ethylene feedstock, these uses as well as utilities used within the various process units are included within the system boundary.

Natural resource extraction, transport operations, and crude oil distillation are excluded from the analysis, since impacts are common for all case options explored. If, however, the study had compared diesel from crude oil with diesel from renewable feedstocks, resource extraction and transportation steps would need inclusion, due to basic differences in the raw materials and their extraction.

Generation of electricity, steam, and cooling water, and use of gas in fired heaters are included. However, distribution of these utilities from utility generation plants to the process units is excluded. Other operations excluded include catalyst manufacturing and disposal, sour water treatment, combustion of light ends, construction of processing units, and the use of recovered sulphur. Exclusions were based primarily on judgements regarding the relative magnitude of impacts.

### 11.4.3 *Inventory Data*

Table 11.2 summarises the inventory data for the overall processing system, comprising the hydrotreater unit, dedicated hydrogen plant (SRU), amine unit, and the sulphur recovery unit. Data were based on flow sheet mass and energy balance calculations, supplemented by estimates of fugitive emission over the four plants. Inventory data are expressed in tonnes per barrel of straight run gas oil (SRGO) fed to the hydrotreater. Hydrotreater capacity was selected as 15,000 bbl/day of SRGO treated.

Inventory data in Table 11.2 show that the hydrotreater is generally the major source of resource consumption and waste emissions. These

Table 11.2  System inventory data (t/bbl SRGO) (Burgess & Brennan, 2001)

| | Hydrotreatert | Hydrogen production | Amine unit | Sulphur recovery | Total |
|---|---|---|---|---|---|
| **Resource used** | | | | | |
| Coal | $3.83 \times 10^{-4}$ | $2.61 \times 10^{-4}$ | $1.04 \times 10^{-4}$ | $1.22 \times 10^{-4}$ | $8.70 \times 10^{-4}$ |
| Water | $4.36 \times 10^{-2}$ | $2.52 \times 10^{-2}$ | $1.70 \times 10^{-2}$ | $1.09 \times 10^{-3}$ | $8.69 \times 10^{-2}$ |
| Natural gas | $1.63 \times 10^{-3}$ | $2.36 \times 10^{-3}$ | $5.53 \times 10^{-4}$ | $-3.72 \times 10^{-4}$ | $4.17 \times 10^{-3}$ |
| **Waste generated** | | | | | |
| $CO_2$ | $5.94 \times 10^{-3}$ | $5.21 \times 10^{-3}$ | $1.94 \times 10^{-3}$ | $-8.89 \times 10^{-4}$ | $1.22 \times 10^{-2}$ |
| $SO_2$ | $4.01 \times 10^{-6}$ | $2.68 \times 10^{-6}$ | $1.10 \times 10^{-6}$ | $1.16 \times 10^{-5}$ | $1.94 \times 10^{-5}$ |
| $NO_x$ | $7.08 \times 10^{-6}$ | $2.85 \times 10^{-6}$ | $2.25 \times 10^{-6}$ | $-5.06 \times 10^{-7}$ | $1.17 \times 10^{-5}$ |
| Solid waste | $5.55 \times 10^{-7}$ | $2.90 \times 10^{-6}$ | $0$ | $5.59 \times 10^{-6}$ | $9.05 \times 10^{-6}$ |
| $H_2S$ | $1.52 \times 10^{-6}$ | $0$ | $1.47 \times 10^{-6}$ | $4.71 \times 10^{-6}$ | $7.70 \times 10^{-6}$ |
| HCs | $2.62 \times 10^{-5}$ | $6.70 \times 10^{-8}$ | $2.56 \times 10^{-6}$ | $7.76 \times 10^{-7}$ | $2.96 \times 10^{-5}$ |
| $CH_4$ | $8.46 \times 10^{-7}$ | $5.20 \times 10^{-7}$ | $3.15 \times 10^{-7}$ | $1.10 \times 10^{-8}$ | $1.69 \times 10^{-6}$ |

Table 11.3   Utilities contribution to life cycle inventories (Burgess, 1999)

| Input/Output | Total inventory ($t$/bbl SRGO) | % contribution from utilities to total inventory |
|---|---|---|
| Resource used | | |
| Coal | $8.70 \times 10^{-4}$ | 100 |
| Water | $8.69 \times 10^{-2}$ | 80 |
| Natural gas | $4.17 \times 10^{-3}$ | 57 |
| Waste generated | | |
| $CO_2$ | $1.22 \times 10^{-2}$ | 78 |
| $SO_2$ | $1.94 \times 10^{-5}$ | 44 |
| $NO_x$ | $1.17 \times 10^{-5}$ | 83 |
| Solid waste | $9.05 \times 10^{-6}$ | Minor |
| $H_2S$ | $7.70 \times 10^{-6}$ | Negligible |
| HCs ($C_2^+$) | $2.96 \times 10^{-5}$ | Negligible |
| $CH_4$ | $1.69 \times 10^{-6}$ | Negligible |

effects stem from the high electricity, cooling water, steam, and natural gas requirements for the hydrotreater unit, reflecting the elevated reactor temperature and pressure.

Steam is a by-product for both the hydrogen and sulphur recovery units; it is assumed that there is demand for this steam elsewhere in the refinery. Hence, consumptions and emissions associated with alternative means of steam generation have been credited to these units, leading to some inventories being represented as negative numbers. Table 11.3 shows the high proportion of inventory which can be attributed to utilities use as distinct from process streams. This highlights the importance of identifying and reporting impacts derived from utilities.

Figure 11.5 shows the results of inventory estimates for the hydrotreater unit. Sulphur content in the gas oil is reduced from 2.19 wt % in the feed to 0.05 wt % in the product. Raw material and product streams, utilities, spent catalyst, and fugitive emissions are shown. Raw materials consumed and effluents generated as a result of utilities consumed are also shown. All quantities are per bbl SRGO feed, selected as the *functional unit*. Similar balances have been performed on the hydrogen, amine, and sulphur recovery units in the processing system.

Scaling factors were applied to input and output data for each processing unit so that inventories can be related to the functional unit of 1 barrel of gas oil feed. For the hydrogen plant, inventory data were based on a plant capacity of 80 tonnes per day (tpd). Since only 9.2 tpd hydrogen are required

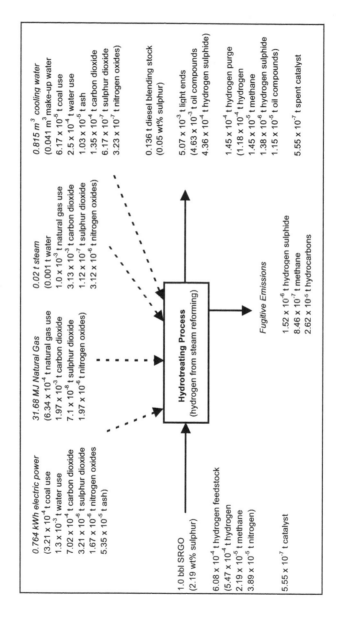

**Figure 11.5** Inventory data for hydrotreating process (Burgess & Brennan, 2001).

by the hydrotreating unit, inventory values calculated for the hydrogen plant were scaled by a multiplier of 0.115.

Where hydrogen is supplied from a CRU, hydrogen is a by-product from this unit. If CRU impacts are allocated to the hydrogen stream on a mass flow basis, a scaling factor of 0.017 results. While this factor is small, it reflects the CRU's dominant function in a refinery as an octane enhancer of gasoline, rather than a hydrogen producer.

### 11.4.4 *Impact Assessment*

The impact assessment consisted of classification, characterisation, and normalisation steps. Impact categories judged important in this study included

- resource depletion
- global warming
- acidification
- photochemical oxidant potential
- nutrification
- human toxicity
- solid waste generation.

Table 11.4 shows the classification and characterisation for 1 bbl of SRGO treated with hydrogen from the dedicated plant. Resource depletion was not rigorously assessed on the basis that

- assigning coal to electricity, and natural gas to both fired heaters and steam generation was to some extent arbitrary
- resource scarcity would be screened out in the normalisation step.

*Normalisation* was performed on impacts from emissions by dividing each score by the total score for that region or for the world. The normalised acidification score for the world for example equals

$$\frac{\text{acidification score for system [kg SO}_2/\text{bbl SRGO]}}{\text{annual global acidification score [kg SO}_2/\text{year]}} \qquad (11.1)$$

The units of normalised impact scores are thus 'years/bbl SRGO'. Estimates of global and Australian data used in normalising impacts are shown in Table 11.5.

Mass inventories for coal and natural gas were converted to their energy equivalents and summed, and classified as normalised 'energy depletion'. Normalised *global* impacts are shown in Fig. 11.6. This is a typical graphical

Table 11.4  Classification and characterisation for 1 bbl SRGO treated ($H_2$ from dedicated plant) (Burgess and Brennan, 2001)

| | Resources used | | | Wastes generated | | | | | | | |
|---|---|---|---|---|---|---|---|---|---|---|---|
| | Coal | Natural gas | Water | $NO_x$ | $CO_2$ | $SO_2$ | Solid waste | $H_2S$ | $C_2^+$ Hydrocarbon | $CH_4$ | Total |
| **Inventory (kg)** | 0.87 | 4.17 | 86.9 | 1.2E-02 | 12.2 | 1.94E-02 | 9.05E-03 | 7.70E-03 | 2.96E-02 | 1.69E-03 | |
| **Equivalency factors** | | | | | | | | | | | |
| Resource depletion | 1 | 1 | 1 | | | | | | | | |
| Global warm, kg $CO_2$/kg | | | | | 1 | | | | | 11 | |
| Acidification, kg $SO_2$/kg | | | | 0.7 | | 1 | | | | | |
| Photochemical oxidant formation, kg $C_2H_4$/kg | | | | | | | | | 0.377 | 0.007 | |
| Human toxicity, kg/kg | | | | 0.78 | | 1.2 | | 0.78 | | | |
| Nutrition, kg phosph/kg | | | | 0.13 | | | | | | | |
| Solid waste, kg/kg | | | | | | | 1 | | | | |
| **Impact score** | | | | | | | | | | | **Total** |
| Resource depletion, kg | 0.87 | 4.17 | 86.9 | | | | | | | | 91.9 |
| Global Warming kg $CO_2$ | | | | | 12.20 | | | | | | 12.22 |
| Acidification, kg $SO_2$ | | | | 8.2E-03 | | 1.94E-02 | | | 1.86E-02 | | 0.0276 |
| Photochemical oxidant formation, kg $C_2H_4$ | | | | | | | | | 1.12E-02 | 1.18E-05 | 0.0112 |
| Human toxicity, kg | | | | 9.1E-03 | | 2.33E-02 | | 6.01E-03 | 0.0384 | | |
| Nutrification, kg phosph | | | | 1.5E-03 | | | | | | | 0.00152 |
| Solid waste, kg | | | | | | | 9.05E-03 | | | | 0.00905 |

Table 11.5   Global and Australian normalisation data (Burgess & Brennan, 2001)

| Impact | Global | Source | Australian |
|---|---|---|---|
| Energy depletion (GJ/yr) | $2.35 \times 10^{11}$ | UNEP (1996) | $4.2 \times 10^{9(1)}$ |
| Water use (m³/yr) | — | — | $1.46 \times 10^{10(2)}$ |
| Global warming (kg $CO_2$/yr) | $3.77 \times 10^{13}$ | UNEP (1996) | $4.03 \times 10^{11(3)}$ |
| Acidification (kg $SO_2$/yr) | $2.86 \times 10^{11}$ | UNEP (1996) | $2.94 \times 10^{9(4)}$ |
| Photochemical oxidant formation (kg $C_2H_4$/yr) | $3.74 \times 10^9$ | UNEP (1996) | $8.95 \times 10^{8(4)}$ |
| Human toxicity (kg/yr) | $5.76 \times 10^{11}$ | UNEP (1996) | — |
| Nutrification (kg phosphate/yr) | $7.48 \times 10^{10}$ | UNEP (1996) | — |
| Solid waste (kg/yr) | — | — | $6.6 \times 10^{11(5)}$ |

[1] 1993–1994 data
[2] 1987 data
[3] 1995 data
[4] 1995 data
[5] 1996 data
**Note:** Details of data sources are provided in (Burgess & Brennan, 2001).

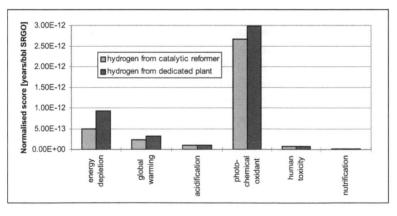

**Figure 11.6**   Environmental scores normalised using global normalisation data (Burgess and Brennan, 2001).

representation of impact profiles, showing normalised impact scores for each impact category. The figure indicates that hydrogen sourced from the dedicated plant results in marginally higher impact than hydrogen sourced from the catalytic reformer. It also indicates that photochemical oxidant potential is the dominant impact, exceeding that of energy depletion and global warming. However, if impacts are normalised *using Australian data*, the dominance of photochemical oxidant potential is reduced, and less than

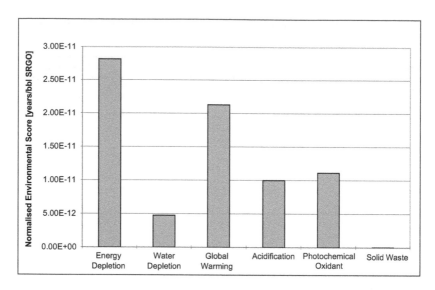

**Figure 11.7** Environmental scores normalised using Australian normalisation data (Burgess and Brennan, 2001).

energy depletion and global warming. Figure 11.7 shows normalised scores for Australia based on hydrogen from the catalytic reformer. The different relativity in impacts is due to higher emission of gases contributing to photochemical oxidant potential in Australia than in Holland, which was the basis for the UNEP estimate of global data.

The profile for energy depletion, global warming, and acidification categories, however, is similar for the global and Australian data. Australian data were not available for human toxicity and nutrification categories. Data for Australia, Holland, United Kingdom, and Western Europe show there is significant variation in man-made volatile organic compound (VOC) emissions for different regions, thus impacting on the normalised score obtained for photochemical oxidant formation. It is of interest that the contributions to photochemical oxidant potential are those from fugitive hydrocarbon emissions. This emphasises the importance of both assessing fugitive emissions in the inventory analysis, and reducing these emissions in plant design and maintenance.

### 11.4.5 *Environmental Burden versus Benefit Comparison*

The environmental benefit of combustion of a lower sulphur diesel can be compared with the environmental burden incurred in the extended

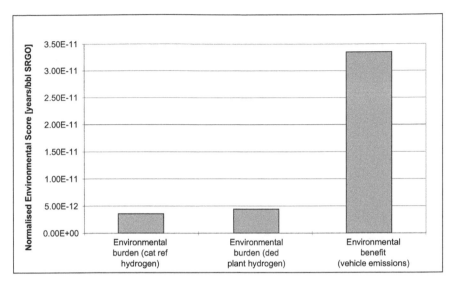

**Figure 11.8** Environmental burden versus benefit of desulphurising: 2.19 wt% to 0.05 wt% S (Burgess & Brennan, 2001).

hydrotreating system. This comparison is illustrated in Fig. 11.8 for the case of reducing sulphur in gas oil from 2.19 wt% to 0.05 wt %, where the benefits clearly outweigh the burdens.

The comparison has been made for the case of a gas oil feedstock of lower sulphur content, where reduction is from 0.16 wt% to 0.05 wt % S (Fig. 11.9). In this case, burdens outweigh the desulphurisation benefits. (Hydrotreating has additional benefits such as reduction in aromatics and in nitrogen which have been ignored in this comparison.)

While the estimated benefit versus burden balance would improve with fewer fugitive emissions and with reduced weighting of photochemical oxidant impact, the balance is clearly less favourable for low sulphur feeds than for high sulphur feeds. This is because elevated pressures and temperatures in the hydrotreater are required independent of the sulphur concentration in the feed.

## 11.4.6 *Conclusions*

This study illustrates how LCA studies can be used for assessing environmental burdens. In the process system studied, the hydrotreater has been demonstrated as the major contributing unit to environmental burdens. Interestingly it was also the major contributor to capital and operating

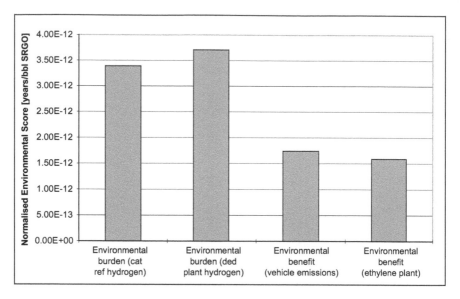

**Figure 11.9**  Environmental burden versus benefit of desulphurising 0.16 wt% to 0.05 wt % S (Burgess & Brennan 2001).

costs, as discussed in Chapter 14. While there were differences in costs and environmental burdens in the two sources of hydrogen explored, the effect of these differences within the overall system was small. The environmental benefit of hydrotreating relative to the environmental burden incurred in hydrotreating is less for lower sulphur feeds.

Aspects of LCA methodology which were evident in the study included

- the importance of utility systems in inventory data
- the potential contribution of fugitive emissions to inventory data
- the sensitivity of normalised impacts to the region under consideration
- the policy for allocation of burdens between product streams.

## Acknowledgements

Case studies are drawn from the work of K. A. Golonka, A. A. Burgess, and J. R. May, their theses and joint publications with D. J. Brennan.

Figures 11.6, 11.7, 11.8, 11.9, 11.10, and 11.11 and Tables 11.2, 11.4, and 11.5 have been reprinted from Burgess and Brennan (2001) with permission from Elsevier. Table 11.1 is adapted from May and Brennan (2003, 2004).

## References

Bently, K. M. and Jelinek, S. F. (1989) NOx Control Technology for Boilers Fired with Natural gas or Oil, Tappi Journal, pp. 123–130, April.

Burgess, A. (1999) *An Environmental and Cost assessment of Gas Oil Desulphurisation*, MEngSc Thesis, Monash University.

Burgess, A. A. and Brennan, D. J. (2001) Desulfurisation of gas oil. A case study in environmental and economic assessment, *J. Clean. Prod.*, 9, 465–472.

Gary, J. H. and Handwerk, G. E. (1994) *Petroleum Refining. Technology and Economics*, 3rd edn, Marcel Dekker, New York.

Golonka, K. A. (1996) *Strategies for Treatment of Smelter Gases Containing Sulphur Dioxide*, PhD Thesis, Monash University.

May, J. (2004) *Sustainability of Electricity Generation Using Australian Fossil Fuels*, PhD Thesis, Monash University.

May, J. R. and Brennan, D. J. (2003) Life cycle assessment of Australian fossil fuel energy options, *Transactions of the Institution of Chemical Engineers*, Vol. 81 Part B, pp. 317–330, and (2004) Vol. 82 Part B, p. 81.

Peters, M. S. and Timmerhaus, K. D. (1991) *Plant Design and Economics for Chemical Engineers*, 4th edn, McGraw-Hill, New York.

# Chapter 12

# Safety Evaluation

## 12.1 Introduction

Safety is often not presented in the sustainability literature as a key sustainability criterion. However, safety is indisputably of vital importance in judging the social merits of a process technology, plant, product, or industry. This is because of the potential threat to life, health, or well-being not only of employees, but also of resident communities near factories and users of industrial products. In conventional sustainability terminology, these three social groups are key stakeholders.

Safety has long been an acknowledged issue for the process industries, particularly in view of the large inventories of hazardous materials processed, stored and transported. The main hazards, derived from the flammable and toxic properties of materials handled, are fire, explosion, and toxic release. These are superimposed on the range of commonly encountered industrial hazards such as

- tripping
- falling objects
- electrical hazards
- asphyxiation
- noise
- working in confined spaces
- health effects from exposure to chemicals.

*Sustainable Process Engineering: Concepts, Strategies, Evaluation, and Implementation*
David Brennan
Copyright © 2013 Pan Stanford Publishing Pte. Ltd.
ISBN 978-981-4316-78-1 (Hardcover), 978-981-4364-22-5 (eBook)
www.panstanford.com

The 1960s saw a strong emergence of safety consciousness within the chemical process industries, spurred on by the development and construction of large single stream plants to produce many commodity chemicals. Large plants brought economies of scale in capital and production costs, but there were downsides. Large inventories of hazardous materials implied substantial safety hazards. Deficiencies in reliability implied economic penalties through production loss, and also potential hazards derived from abnormalities in plant operation. There have also been concerns for public safety arising from transport of hazardous materials from centralised, large-scale plants to off-site downstream processing plants and other customers.

Many internal company initiatives were under way in 1974, when the *Flixboro disaster* spelt out the sentence that many had feared. Release of a large inventory of flashing cyclohexane liquid, with a resulting vapour cloud explosion flattened the caprolactum plant, killing all on-site and damaging nearby houses. This event had a major impact on accelerating the drive towards inherent safety — smaller inventories, safer processing conditions for hazardous materials, and substituting less hazardous process materials. There was also a trend towards building fortress-like control rooms, such that in the event of a major hazard, personnel within control rooms would be safe and records of the incident could be preserved for future learning.

Further judgement was passed in 1984 with the *Bhopal incident* in India — this time involving a massive release of toxic methyl isocyanate, killing or adversely affecting huge numbers of civilians living close to the plant. Again the message of inherent safety was emphasised, with the theme of 'what you don't have can't leak' (Kletz, 1978) having poignant effect. The implications of *safe boundary distances* from processing facilities were also underlined.

In University chemical engineering courses in the early 1960s, little was taught on process safety, except in the context of pressure vessel design and pressure relief provisions. An important message that escaped many undergraduates was the importance of process and plant design being sufficiently robust to failures in plant, equipment, utilities and instrumentation, as well as operator error. It was not until the mid 1970s that hazard and operatability (HAZOP) studies, exploring outcomes of deviations from design intent, became common in the chemical industry. Since then, major developments in safety curriculum have occurred, spurred on by professional society accreditation.

Engineers and scientists within industry are required to address a wide range of situations arising from the diverse nature of process technologies, the materials handled, and the range of temperature, pressure, and concentration conditions encountered. A range of perspectives, knowledge areas, and tools must be adopted including process safety fundamentals and also involving stoichiometry, fluid mechanics, radiant heat transfer, thermodynamics, and kinetics, as well as process and plant design. Process safety topics are addressed both within design texts (e.g., Sinnott & Towler, 2009) and specialist texts (e.g., Crowl & Louvar, 2002), the latter having a strong emphasis on quantitative analysis. Table 12.1 summarises some key issues in process safety evaluation.

## 12.2 Importance of Learning from Accidents, Dangerous Occurrences

Learning from past disasters is a very important part of process safety. These past incidents have usually been thoroughly scrutinised, debated, and reported through legal public enquiries, providing valuable opportunity for study and learning. Learning is instructive in relation to

- the inherent causes of the disaster
- the sequence of events leading to the disaster
- the human shortcomings (whether in knowledge, training, response, management) leading to the disaster
- the resulting impacts on society, the environment and the economy.

Useful references for learning about previous major industrial accidents include Lees (1996), Marshall and Ruhemann (2001), and the IChemE's Loss Prevention Bulletin. The IChemE's Loss Prevention Bulletin, published monthly, not only reports on a wide range of incident types within the process industries, but provides excellent insight into aspects of plant design and the nature of plant operations. Unfortunately there have been industrial accidents which have not had adequate scrutiny because of fortunate escapes from death or injury, small numbers of people hurt, or because the accident has not had a major impact on the wider community. Proper investigation of all dangerous occurrences is imperative both to ensure corrective action and learning to avoid future recurrence.

Structured root cause analysis (Sutton, 2010) can be beneficial in identifying root causes of accidents and 'near misses'. A sequence of events leading to an accident might be identified initially in terms of failures by instruments or by human operator error. However, there may be underlying causes of failure such as procurement of unreliable instruments, or systemic management failures in relation to operating procedures or training. Root cause analysis was an integral part of the investigation into the explosion in

Table 12.1    Check list of some key considerations in process safety evaluation

---

**Flammability Properties of Substances** [1]
Flammability limits
Flash point
Auto-ignition temperature
Heat of combustion

**Ignition Sources**
Common ignition sources in process plants include
- Electric motors
- Open flames, for example, in fired heaters
- Static electricity
- Others

**Fires**
Fire types include pool fires and flash fires. Resulting damage can include
- Radiation damage to people, materials, buildings
- Asphyxiation from envelopment in flames
- Secondary effects such as contributing to BLEVE's (boiling liquid expanding vapour explosions)

**Explosions**
Explosions involve sudden violent release of energy (mechanical or chemical).
Chemical explosions are classified as detonations or deflagrations.

Explosions are sometimes classified as confined or unconfined.

**Resulting damage** from deflagrations include
- Effects of overpressure which decrease in severity with increased distance from the explosion
- Possible secondary damage from missile effects

**Toxicity** [2]
**Acute toxicity** (short-term effects) is distinguished from **chronic toxicity** (long-term effects).
Safe working exposure limits are set for various chemicals to protect plant operating personnel.
For accidental release of toxic material, atmospheric dispersion plays an important role in reducing concentrations within and beyond boundaries of plant site

**Inerting and purging** are important strategies
- For avoiding ingress of air (and hence oxygen) into piping and equipment handling flammable materials
- Removing toxic or flammable materials from piping and equipment prior to isolation for maintenance

**Overpressure and Underpressure**
Subjecting piping or equipment to pressures above or below their intended or design conditions

Table 12.1  *(Contd.)*

---

- Causes, including equipment failure, utility supply failure, operator error
- Consequences, including air ingress (underpressure) and collapse or fracture
- Preventive approaches, including alarms, trips, interlocks
- Methods of pressure relief, including relief valves and busting discs, and safe disposal of released chemicals

**Hazards Associated with Storage**

Atmospheric pressure storage of flammable or toxic liquids — related risks of underpressure, overpressure, fluid leakage and air ingress

Pressurised storage of toxic or flammable liquefied gases under ambient temperature conditions — related risks of fluid release Potential for BLEVE's in pressurised storage of flammable liquefied gases

**Inherent Safety**

- Adoption of new process routes with safer materials and conditions
- Substituting non-toxic for toxic materials
- Reducing inventories of hazardous materials
- Providing less severe conditions, for example, of pressure, temperature, and concentration to minimise potential hazards

**Plant Layout Provisions**

- Separating concentrations of people from hazardous areas
- Separating flammable material areas from ignition sources
- Considering prevailing wind directions

---

Note: [1],[2]For definitions and properties, see Crowl and Louvar (2002). For notes on flammability limits within chemical processes, see Appendix 3 in this chapter.

the Isomerisation Unit at BP refinery in Texas City in March 2005 (Broadribb, 2006).

## 12.3 Life Cycle Issues

Apart from focusing on safety issues within process plants, safety is a key concern in the extraction and purification of resources, in the transport of raw materials, intermediates and products, and in product use. For mineral resources, there is concern for the safety of mining, especially underground mines, from the viewpoints of ground stability, ventilation, and secure routes of access and egress for personnel. Underground coal mines have the dual hazards of ground stability and fire/explosion risks. A dramatic example of potential hazards in oil and gas extraction was the Piper Alpha incident (http://www.fabig.com/Accidents/Piper+Alpha.htm) and (Cullen, 1990). Piper Alpha, a major offshore oil and gas production platform operating in the North Sea, was destroyed by fire and explosion in 1988 resulting in 167 deaths.

Many fuels, chemicals, and other materials with toxic, flammable or otherwise hazardous properties are transported by road, rail, or ship. Risks associated with such transport were discussed in Chapter 9. One of the most damaging road accidents was the so called 'Spanish Campsite Disaster' in 1978. In this disaster a road tanker containing 23 tonnes of liquid propylene burst near a campsite in Spain (close to Tarragona on the east coast), releasing a large vapour cloud which ignited. Some 215 deaths, 67 serious injuries, and widespread property damage were reported (Marshall & Ruhemann, 2001).

Towards the end of a product life cycle there are the issues of safety in product use and ultimate disposal. Many products have implications for safety and the environment when disposed of through incineration or at landfill sites. End of life concerns also apply to numerous facilities and plants, including contaminated soils at petrol stations, contaminated soils and groundwater at used chemical plant sites, landfill gases and ground water contamination at landfill sites, emissions from incinerators, and mothballed nuclear power plants.

## 12.4  Health, Safety, and the Environment

Many companies and industries have aggregated their efforts in health, safety, and the environment. There are important links between these areas encouraging this integrated approach. For example, there are human health, safety and environmental impacts arising from exposure to

- toxic substances
- noise
- excessive temperatures (thermal pollution, burns, damage to materials).

Health, safety, and environmental damages also frequently result from industrial accidents. Examples of environmental damage arising from major industrial accidents include the disasters of the well-publicised Seveso, Chernobyl, and Basel incidents reported under Section 12.5.

There are often common precursors to safety and environmental accidents including

- abnormality in operation
- failure of hardware such as instruments, valves, equipment items

- failure of utilities such as electricity, cooling water, steam
- human failure, whether of operator, engineer, or manager.

There are also common outcomes of safety and environmental incidents including

- societal impacts — personal, family and community
- impacts on the image of a company or industry sector
- interruption to production or service provision
- loss of product quality
- financial losses (arising from loss of production, loss of facility, damage to plant, increased insurance premiums).

## 12.5 Examples of Safety Incidents in the Process Industries Involving Environmental Damage

Some safety incidents which also resulted in environmental damage are now briefly discussed.

### A. Seveso (15 miles from Milan, Italy) — 1976 [Ref. Appendix 2 in Lees 1996]

Rupture of a bursting disc on a chemical reactor occurred at a chemical works: a white cloud drifted from the chemical works and material was deposited downwind. Among substances deposited was a very small amount of TCDD (2,3,7,8-tetrachlorodibenzoparadioxin), one of the most toxic chemicals known.

There followed a period of confusion due to lack of communication between the company and the authorities, and the latter's inexperience in dealing with this kind of situation. In the next few days in the affected area, animals died and people fell ill. A partial, belated evacuation took place. In the immediate aftermath there were no deaths directly attributable to TCDD but a number of pregnant women who had been exposed suffered miscarriages.

### B. Chernobyl (Ukraine) — 1986 [Ref. Appendix 2 in Lees 1996]

On Monday, 28 April 1986, a worker at a nuclear power station in Sweden put his foot in a radiation detector for a routine check and registered a high reading. The power station staff thought they had had a radioactive release from their plant and the alarm was raised. In fact an accident had occurred 800 miles from Sweden at Chernobyl on Saturday, 26 April (1.24 am). An experiment to check the use of a turbine during rundown as an emergency power supply for the reactor went catastrophically wrong. There was a power surge in the reactor, the coolant tubes burst and a series of explosions rent the concrete containment. The graphite moderator caught fire and burnt sending out a plume of radioactive material. Emergency measures to put out the fire and stop the release were not effective until 6 May, some 10 days later.

Of those on-site at the time, two died on-site and a further 29 died in hospital over the next few weeks. All others on-site were affected to varying degrees. There have also been ongoing health problems experienced by the local population. There were mitigating features in that at the time of the initial release of radioactive material there was dry weather, no wind, and it was night with most people indoors.

### C. Basel (Switzerland) — Warehouse fire (1986)

A large fire occurred in a warehouse used by Sandoz (a Swiss chemical manufacturing company). The fire generated heavy smoke containing offensive materials, and the local population was warned to take refuge indoors. More serious consequences resulted from the use of water by firefighters, which subsequently drained into the River Rhine, carrying toxic chemicals. Severe ecological damage was caused to the river over a distance of 250 km, including the death of a large number of fish and eels.

### D. Coode Island (Melbourne, Australia) — 1991

Approximately 8.5 million litres of chemicals burned following a fire and failure of an acrylonitrile storage tank at Victoria's largest toxic chemical storage facility. Huge clouds of toxic smoke and fumes threatened nearby suburbs, but fortunately were largely dispersed by high winds in the area. More than 250 people were evacuated from nearby factories and ships. Of the 150 firefighters who fought the inferno, 2 were injured. A total of 14 storage tanks containing about 600,000 L of chemical each were destroyed and 230,000 litres of fire-suppressing foam were used against the blaze. Damage to the facility and clean-up costs were estimated at over $20 m (Australian $ 1991) (Ref. www.ema.gov.au/).

### E. Oil rig accidents (2010, 2009)

An explosion and fire (April 2010) on the Deepwater Horizon drilling rig was reported by news media to have caused 11 deaths of crew on the rig and subsequent oil leak in the Gulf of Mexico threatening the coast of Louisiana, Texas and the Mississippi estuary in the United States. Estimates of leakage rates reported have ranged from 1000 to 8000 barrels a day.

A separate explosion and fire (August 2009) and associated oil leak of 400 to 3000 barrels per day were reported by news media in the Montana oil field, west of Darwin, Australia.

**F. Oil release from shipping accidents**

Cases of sea pollution with serious consequences have occurred with the grounding of tankers on coasts. There are numerous examples, including (1967) Torrey Canyon carrying 114,000 tonnes of crude oil on the coast of Cornwall; (1978) Amoco Cadiz lost its steering and ran aground on Brittany coast; (1989) Exxon Valdez ran aground at Alaska rupturing 8 cargo tanks and releasing some 258,000 barrels of oil.

## 12.6  Accident Prevention

There are a number of approaches to accident prevention. Design reviews are held at key stages of project development and process design for new plants, and for proposed modifications to operating plants. These reviews aim to ensure that systems, procedures and equipment are in place to prevent an initiating failure, and that adequate and reliable lines of defence are available in the event of such a failure.

A key step in early stages of project development is hazard identification, assisted by check lists. Some examples of such check lists can be found in Sinnott and Towler (2009) Chapter 9, in conjunction with the outline of the Dow Fire and Explosion Index and in Crowl and Louvar (2002) Chapter 10, which deals specifically with hazard identification. It is important in the early stages of process development or process design to strive, where possible, to use process materials and conditions which will favour both safety and the environment.

### 12.6.1  *The HAZOP Approach*

*The HAZOP approach* developed for project safety seeks to identify deviations from intent. A 'what if' scenario is systematically explored for a complete range of deviations for each intended condition of a process stream, equipment item or section of pipe. This explores causes, consequences, and corrective action, whether at design or operational level.

Examples of deviation applied to level, flow, pressure, temperature, composition include

> HIGH
> LOW
> ZERO
> REVERSE (for FLOW)

NEGATIVE (for PRESSURE)
CONTAMINANT (for COMPOSITION).

A similar approach is explored through event scenarios, for example, what happens if there is

- failure of a key utility such as steam, electricity, cooling water, or nitrogen
- spillage of a hazardous substance
- leakage, either into or out of a process system
- accident in transit (ship, train, road tanker) or in loading/unloading operation
- material impurity whether
  - ▶ in a raw material to process
  - ▶ in a consignment of waste for landfill
  - ▶ resulting from a specific error by an operator.

The HAZOP study is normally performed on piping and instrumentation diagrams. A key element of success of the HAZOP study is the extent and diversity of experience that can be brought to bear on the study through team membership. It is important to have design, operational, and maintenance experience within team membership, and good team management to secure useful outcomes.

A desirable output from the HAZOP study is a plant which is robust to failure tolerance in plant, equipment, instrumentation, utilities, as well as its management systems and people. Some key aspects of failure tolerance are summarised in Appendix 1 of this chapter.

HAZOP studies were specifically developed for safety reviews but their principles can be similarly applied to environmental accidents and releases.

## 12.7 Techniques for Investigating Probability of Major Incidents

Techniques are used for exploring sequences of events which can develop from a particular fault occurring, and also sequences of events which can lead to a specified undesirable outcome. These include

- the event tree, which starts with an event (e.g., loss of cooling water) and examines possible consequences, outcomes, and their probability.
- the fault tree, which starts with a serious end result (e.g., an explosion or over pressuring a vessel), referred to as a 'top event',

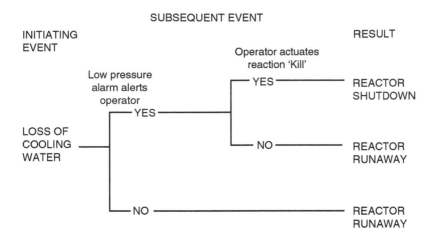

**Figure 12.1**   Example of event tree initiated by cooling water failure.

and examines possible sequences of events leading to that end result. From estimates of the probabilities of contributing causes, the probability of the end result is then estimated.

Simple examples are provided of an event tree (Fig. 12.1) and a fault tree (Fig. 12.2). A useful explanation of event and fault trees, supported by examples, is provided by Crawley and Tyler (2003).

Figure 12.1 shows some possible events following the loss of cooling water to a batch reactor undergoing an exothermic reaction. A runaway reaction is possible and in the most serious case can result in loss of contents through pressure relief. One remedial action is to add a reaction stopper (see discussion in Chapter 4 under Section 4.16).

Figure 12.2 shows some conditions leading to complete loss of pumping capacity for cooling water. In this case, cooling water is pumped from a cooling tower using two centrifugal pumps, each operating continuously and each sized to provide 50% of required cooling water capacity. One pump (Pump A) is driven by a steam turbine, while the other pump (Pump B) is electrically driven.

## 12.8  Risk Assessment

Risk assessment involves a quantitative evaluation of whether the risks associated with any hazard meet established criteria of acceptability. Risk

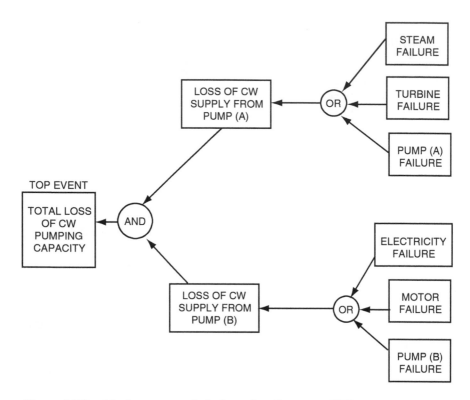

**Figure 12.2** A fault tree example for loss of cooling water (CW) pumping capacity.

(or probability) of death from an incident can be estimated and compared, for example, with fatality statistics associated with travel or recreation activities.

Risk is commonly interpreted as the combination of

- consequence (severity)
- likelihood (frequency).

Risk matrices, providing a tabular representation of consequence and likelihood, are commonly used during or after hazard identification studies, to screen hazards or to conduct a simple risk analysis. A useful discussion of the use of risk matrices is provided by Middleton and Franks (2001).

## 12.9 Preventive Approaches

A number of preventive approaches can be used to avoid major accidents or minimise their consequences. These include

- inherent safety initiatives involving
  - ► new process routes with safer materials and conditions
  - ► substitution of hazardous materials with non-hazardous materials
  - ► reduced inventories of hazardous materials
  - ► less severe conditions of pressure, temperature, concentration
- inerting and purging where flammable or toxic materials are used
- removing potential ignition sources
- providing a plant layout which
  - ► limits concentrations of people in hazardous areas
  - ► takes account of the prevailing wind direction
  - ► separates flammable materials from ignition sources.

Principles of inherent safety are summarised in further detail in Appendix 2 of this chapter. A number of indices have been proposed to provide a quantitative assessment of inherent safety. A well known and widely used example is the Dow Fire and Explosion Index. Such indices can be used for identifying the most hazardous steps in a process or the most hazardous section of a process plant. Introductions to the use of the Dow Fire and Explosion index are provided by Crawley and Tyler (2003) and by Sinnott and Towler (2009).

The application of the Dow Fire and Explosion index in evaluating inherent safety merits has been discussed within the framework of a sustainability assessment of integrated gasification and combined cycle power generation (IGCC) using black coal (Falcke, Hoadley, Brennan, & Sinclair, 2010). A separate study involving application of inherent safety principles to an onshore gas plant has been reported by Pickering and Brennan (2005).

## 12.10  Transport and Storage of Chemicals

There are environmental and safety hazards associated with both storage and transport of many chemicals of commerce. The process industries have grown up with such hazards, but increasingly there is community pressure to reduce or remove them. There have been relatively recent arguments for a shift in strategy from manufacture of hazardous chemicals in large plants with subsequent transport to the customer, towards a strategy of smaller scale manufacturing plants located at various points of use. The supply and manufacture of chlorine and hydrogen cyanide have been cited as examples. This strategy shift is discussed further in Chapter 15 in the context of industrial planning.

Industrial ecology initiatives, involving exchange of surplus waste and energy streams between plants on the one site or between plants on separate sites in an industrial zone, should be supported by thorough risk assessments. Operational responsibilities of participating plants, including communications between operating personnel of participating plants, should be agreed, clearly defined, and understood.

The damage potential associated with the transport and storage of hazardous materials was identified in Chapter 9.

## 12.11 Government Legislation

It is essential to keep abreast of developments in government legislation dealing with safety issues. A useful review on developments in Britain and Europe is published in the IChemE journal *Process Safety and Environmental Protection*.

An example of government legislation in Victoria, Australia is the Occupational Health and Safety regulations for major hazard facilities which came into effect in 2000. A major hazard facility is a plant which stores, handles or processes quantities of dangerous chemicals or products above a defined threshold. Typically, such facilities include petroleum refineries, chemical and gas processing plants, and storage and distribution sites for hazardous materials. Under this legislation, operating companies are required to

- establish and implement a safety management system;
- identify hazards and resulting incidents, and assess associated risks;
- adopt control measures to eliminate or reduce risk as far as practicable;
- prepare emergency plans in conjunction with local emergency services and the wider community;
- prepare a safety case and apply for a license.

A safety case in this context is a written document outlining measures to prevent a major incident and to deal with a situation which might result from such an incident. Much of the requirements of this legislation is being adopted by other Australian states and much of it can be expected to be applied in the future to smaller plants and facilities. Details of the Victorian safety legislation and related issues and publications can be found on the web at http://www.worksafe.vic.gov.au/.

*Emergency response training* within a plant or company, which involves key groups from the external community as well as company employees, is an important aspect of preparedness for dealing with hazardous events.

There is legislation in most countries covering the transport and storage of hazardous goods, including responsibilities of employers to employees. One important aspect is the need to provide safety data sheets so that employees know the hazards they are dealing with and the necessary precautions to be taken.

## 12.12 Concluding Remarks

The process industries are inherently hazardous because of the toxic and flammable properties of many of the materials handled, the large scale of operations, and the sizeable inventories of materials within plants and storage and distribution systems. The safety and wellbeing of employees and people living near process plants is of major importance in the context of concern for stakeholders under the sustainability ethos.

This concern extends over the complete life cycle of products from extraction of basic raw materials and fuels, the various materials distribution systems, processing and manufacturing, through to product use. The concern also extends over the complete life cycle of plants including construction, operation, and decommissioning.

In this chapter we have outlined some basic principles associated with the identification and prevention of hazards, and introduced some key approaches to risk assessment and reduction, supported by useful references for further reading.

## References

Broadribb, M. P. (2006) 'Lessons from Texas City. A Case History,' Loss Prevention Bulletin, *IChemE.*, 192, 3–12.

Crawley, F. K. and Tyler, B. (2003) *Hazard Identification methods*. European safety center, IChemE, Rugby, England.

Crowl, D. A. and Louvar, J. F. (2002) *Chemical Process Safety*, 2nd edn, Prentice Hall, Upper Saddle River, NJ.

Cullen, Hon. Lord (1990) *The Public Enquiry into the Piper Alpha Disaster*, HMSO, London, England.

Falcke, T. J., Hoadley, F. A., Brennan D. J., and Sinclair, S. E. (2011) The sustainability of clean coal technology: IGCC with/without CCS, *Process Safe. Environ. Protect.*, doi: 10.1016/j.psep.2010.08.002.

Institution of Chemical Engineers (IChemE) (regular publications)
a. Loss Prevention Bulletin.
b. Process Safety and Environmental Protection (Transactions B of IChemE)

Kletz, T. A. (1978) What you don't have can't leak, *Chem. Ind.*, 9, 287–292.

Kletz, T. A. (1999) *HAZOP and HAZAN. Identifying and Assessing Process Industry Hazards*, IChemE, Rugby, England [plus other titles by the same author].

Lees F. P. (1996) *Loss Prevention in the Process Industries: Hazard Identification, Assessment and Control*, 3 volumes, 2nd edn, Butterworth-Heinemann, Boston.

Marshall, V. and Ruhemann, S. (2001) *Fundamentals of Chemical Process Safety*, IChemE, Rugby, England.

Middleton, M. and Franks, A. (2001) Using risk matrices. *The Chemical Engineer (tce).*, September, 34–37.

Pickering, E. H. and Brennan, D. J. (2005) *Application of Inherent Safety Techniques to an Onshore Gas Plant*, Proceedings of Chemeca, Brisbane, Australia.

Sinnott, R. and Towler, G. (2009) *Chemical Engineering Design*, Elsevier, Oxford, England.

Sutton, I. (2010) Incident investigation and root cause analysis, in *Process Risk and Reliability Management* (Chapter 12), Elsevier, Oxford, England.

### Additional Reading

Carson, P. A. and Mumford, C. J. (2002) *Hazardous Chemicals Handbook*, 2nd edn, Butterworth, Oxford, England.

Lees, F. P. and Mannan, S. (eds.) (2005) *Loss Prevention in the Process Industries*, 3 volumes, 3rd edn, Elsevier Butterworth-Heinemann, Amsterdam, the Netherlands.

## Appendix 1. Failure Tolerance

A check list is suggested for use in conjunction with HAZOP approaches to consider the robustness of process plant design:

- process operators
  - ▶ fatigue
  - ▶ lack of training
  - ▶ peak demand, including alarm response
- process Equipment

▶ leakage
▶ corrosion failure

• piping

▶ leakage
▶ corrosion failure

• valves, control valves

▶ valves actuated to close but allowing fluid to pass
▶ valves actuated to open but remaining closed
▶ pressure relief devices leaking when not engaged and failing to relieve properly when engaged

• instruments

▶ inaccurate readings
▶ blocked impulse lines

• utility systems failing to perform reliably or at all

▶ electricity
▶ cooling water
▶ steam
▶ fire water
▶ inert gas for purging.

## Appendix 2. Principles of Inherent Safety (Pickering and Brennan, 2005)

The inherently safer design approach is to eliminate or reduce the hazard by changing the process itself, rather than by adding on additional safety devices and layers of protection. Approaches to the design of inherently safer processes and plants have been grouped into several major categories. The terminology in describing these approaches differs around the world. Regardless of the terminology used, there are essentially four methods of inherently safer design (ISD):

**Substitute:** Replace a material with a less hazardous substance. It can involve the complete elimination of a hazardous material. It is applicable to materials, process routes, and process equipment. Using a safer material in place of a hazardous one decreases the need for added on protective equipment and thus decreases plant cost and complexity.

**Minimise:** Use smaller quantities of hazardous substances, such that if a release occurs, the impact is minimal. This strategy applies when the hazardous material cannot be eliminated. Inventories can often be reduced in almost all unit operations, as well as in storage. This reduction can also bring about reduction in cost, where less material needs smaller vessels, structures, and foundations.

**Simplify:** Design facilities which eliminate unnecessary complexity, make operating errors less likely, and which are robust to operating errors. Simpler plants are usually cheaper and easier to operate and maintain.

**Moderate:** Use less hazardous conditions, a less hazardous form of a material, or facilities which minimise the impact of a release of hazardous material or energy. Moderate embodies the concepts of attenuation and limitation of effects.

## Appendix 3. Some Notes on Flammability Limits

Flammability limits are normally reported for ambient temperature and pressure. However, it is important to recognise that these limits depend on temperature and pressure. This has implications for chemical reactions occurring at elevated temperatures and pressures. Some important industrial examples, all involving reaction at elevated temperatures are

- methanol oxidation to formaldehyde
- methane/ammonia oxidation to hydrogen cyanide
- ammonia oxidation to nitric oxide (precursor to nitric acid).

Crowl and Louvar (2002) give empirical equations for temperature and pressure effects on upper and lower flammability limits.

# Chapter 13

# Assessment of Costs and Economics

## 13.1 Introduction

In this chapter, we explore key economic factors influencing the viability of capital investment in process industry projects. Many reported case studies in completed cleaner production projects have claimed win/win benefits, that is,

- improved environmental performance = WIN
- reduction in costs = WIN.

Economic benefits of cleaner production projects can derive from a number of sources including

- reduced raw materials consumption
- reduced utilities or energy consumption
- reductions in waste emissions, leading to reduced treatment or disposal costs
- increased product sales revenue derived from increased production capacity achieved as a by-product of fundamental technological change.

However, the related capital costs and any operating cost penalties must also be identified and weighed up in the investment proposal.

Capital funds for investing in modification to process plants have to be found from available sources. Capital resources within companies are often

*Sustainable Process Engineering: Concepts, Strategies, Evaluation, and Implementation*
David Brennan
Copyright © 2013 Pan Stanford Publishing Pte. Ltd.
ISBN 978-981-4316-78-1 (Hardcover), 978-981-4364-22-5 (eBook)
www.panstanford.com

limited and must be allocated carefully and responsibly. Even when capital resources are available, the capital investment may not show satisfactory profitability (or return on investment). Possible causes of unsatisfactory profitability include

- finite economic life of the project under investment
- risks associated with the project, whether business or technology related
- the discount rate applied to cash flows reflecting the cost of finance for the project
- time required to shut down an operating plant for modifications, commissioning and start-up; consequent loss of operation and associated income
- time required to develop and prove a technology prior to investment
- costs of purchasing technology
- uncertainty in engineering for a movable target of progressively tightening environmental legislation over the projected life of the plant.

## 13.2  Investment Projects

Capital investment projects may be entirely new production ventures or modifications to existing plants. Examples of possible projects include

- a new power station (large or small)
- a new plant to produce a new cleaner product
- a new plant using cleaner technology to produce an established product
- provision of a waste heat boiler in an existing plant to recover heat energy
- modification to an existing plant to treat an effluent stream.

All projects involve new capital expenditure and involve distinct time spans for

- technology development or purchase
- project planning including selection of site, technology, plant size, and investment timing
- plant design and construction
- plant commissioning and start-up
- plant operation.

It is essential to have an understanding of these activities and their time spans and of key contributions to capital and operating costs, revenues, and project profitability.

## 13.3 Capital Requirements and Sources

Capital may be regarded as accumulated wealth. A company invests capital in the expectation of greater wealth or well-being in the future. For process industry projects, capital investment is required for the following.

- **Land** for the process plant and its essential environs. Investment into land is *recoverable*, and unless the land is degraded, is non-depreciating.
- **Fixed capital into equipment, plant, and buildings** represents the major portion of capital investment into process industry projects, and is essentially *non-recoverable*. Even though equipment may be sold at the end of project life, the resulting capital recovered from the sale is a very small proportion of the total initial investment.
- **Working capital** is investment into *inventories* (or stocks) of *raw materials* and *products*, essential to keep the plant operating continuously and maintaining product supply to customers, as well as the *trading balance* between *debtors* and *creditors*, essential to keep a business operating. Since stocks can potentially be sold into markets and debts and credits settled, working capital is essentially *recoverable*.

There are two main *sources of capital funds*:

- **equity** including
  - ▶ shareholders subscriptions
  - ▶ retained earnings from profitable operations
- **loan sources,** for example
  - ▶ banks
  - ▶ superannuation funds.

Principal and interest must be paid on loans. Interest paid on loans is tax deductible. Equity funds have value through opportunity cost (potential for alternative investments), and shareholders expect a return on their funds. Capital funds for most projects are a mix of equity and loan funds.

## 13.4 Fixed Capital Costs

Fixed capital costs are incurred both for inside and outside battery limits investment. The term 'battery limits' is commonly used to denote the geographic boundary around the main processing plant. *Inside battery limits capital cost* is the capital cost of the processing plant which is dependent on its feedstock, product, manufacturing technology, and production capacity.

*Outside battery limits capital costs* (also sometimes termed 'off-sites' capital) include three main categories:

- storage facilities for raw materials and products
- utilities generation plant, for example, steam boilers, cooling towers
- buildings (including warehouses, laboratories, workshops, offices).

The magnitude of outside battery limits costs depends on

- the means of moving process materials and products to and from the site, often referred to as 'logistics', which influences the required storage capacity for raw materials and products; for example, infrequent shipping of materials can require extensive storage of materials
- the means of utility supply, in particular whether the utility is generated on site or purchased from an external source
- the extent, purpose and characteristics of buildings provided.

**Plant** and **equipment costs** are a function of plant production capacity, process complexity, equipment design characteristics, and choice of construction materials. Plant and equipment costs can ultimately be broken down into costs of

- *materials* (steel, concrete, exotic metals/alloys, refractories, etc.) and
- *labour* (to fabricate equipment, for piping and equipment installation, pouring foundations, designing and managing the project).

This breakdown is useful both in making and checking cost estimates.

### 13.4.1 *Approximate Estimates of Plant*

Costs can be estimated from historic costs (either achieved or estimated) of plants having similar feedstock, product, and processing technology. Allowance must be made for effects of

- plant production capacity
- site location
- inflation of costs from the time of construction or previous estimate to the time of the current estimate.

### Capacity dependence

Capacity dependence is often expressed as the simplified relationship

$$I = kQ^b \tag{13.1}$$

where

$I$ = capital investment (e.g., in $ million)

$Q$ = plant production capacity (e.g., in tonnes of product per day)

$k$ = proportionality factor

$b$ = exponent

The value of exponent $b$ typically approximates 0.5–0.6 for *single stream* plants, and 0.8–0.9 for *parallel stream* plants. *Single stream plants* involve one equipment item (e.g., 1 reactor, 1 distillation column) at each stage of the process and achieve larger capacity by increased equipment sizes. Single stream plants are adopted, for example, for most petroleum refinery units and sulphuric acid plants. *Parallel stream plants* involve multiple equipment items at each processing stage (e.g., 6 reactors, 8 filters, 5 dryers). They reflect cases where the required capacity is too large for a single equipment item to do the task, and are typical of many mineral processing plants such as alumina refineries and aluminium smelters. Where plants adopt both single and parallel stream processing within the overall plant (e.g., multiple reactors with single stream product purification), the exponent lies between the two extremes; the exponent can be weighted according to the relative amounts of capital expended on single and parallel streamed plant.

### Location

Plant costs depend on the cost of labour (including construction labour and design, construction, and management overheads) and the cost of freighting materials and equipment to site. Plant costs may also depend on climate and terrain at the construction site; for example, designing for hurricanes will require higher construction standards while some sites will be less prone to flooding. Labour costs per person employed vary between countries, for example, labour rates have been higher in the United States and Australia than in India and Indonesia. Costs also vary for different locations within a particular country; for example, remote sites (e.g., North West Australia) imply higher labour and freight costs. Labour productivity also influences total cost of labour. Labour productivity varies between countries and over time, and will reflect the skills of labour and management as well as site working conditions.

**Time**

Plant cost inflation indices account for increases over time in labour, material, and equipment costs incurred in building process plants. These indices may be detailed in composition (e.g., Chemical Engineering Plant Cost Index published in the US-based journal *Chemical Engineering* which is an aggregate of many contributing costs appropriately weighted), or a simple amalgamation of labour and materials cost indices. In Australia, detailed composite indices have been unavailable so that simple two-component indices based on Australian Bureau of Statistics (ABS) statistics must be used. Average weekly earnings (for labour), and producer price index (for materials) can be used with a 50% weighting for each. Note that plant costs escalate at different rates to the consumer price index (CPI) which is a composite index designed to measure the cost of living. The CPI encompasses the cost of food, housing, travel, entertainment, health, education, and so on. Plant cost indices for Australia estimated by the author from government statistics are appended to this chapter.

## 13.4.2 *Equipment Costs*

Costs of specific equipment items, for example, pressure vessels, heat exchangers, pumps, depend on

- capacity or size parameter, for example,
  - ▶ volume for a vessel
  - ▶ surface area for a heat exchanger
  - ▶ volumetric flow rate and developed head for a pump
- design pressure and temperature
- materials of construction.

This dependence is widely used as a basis for equipment cost correlations.

There are published data for equipment costs within design texts (e.g., Peters, Timmerhaus, & West, 2003; Sinnott & Towler, 2009) and also within commercial computer packages. It is important to recognise that the purchased cost of equipment, as well as dependent on equipment design, depends on a range of business-related factors. These factors include the business climate at the time of purchase, the extent of the order (discounts may apply for multiple items), profit margins of equipment suppliers, and special client demands regarding delivery times, design or fabricating standards. These influences lead to considerable scatter in data for equipment costs; this scatter must be borne in mind when consulting equipment data bases. Equipment cost data must also be adjusted for inflation.

### 13.4.3  *Contributions to Plant Costs*

Plant costs include the cost of purchased equipment, but also the cost of

- site preparation, roads, drainage and associated civil works
- foundations and structures
- equipment installation
- any process buildings
- piping
- electrics
- instrumentation and
- overheads for design, construction, and project management.

The term 'overheads' is widely used to denote costs which are not restricted to a particular cost category.

Plant costs are typically 3 to 4 times the total cost of purchased equipment. The ratio of *plant cost to purchased equipment cost* is frequently termed a *'Lang' factor* (after Lang, who first observed the generalisation). The *direct plant cost*, excluding *overheads for design, construction,* and *project management*, is typically 2.7 times the purchased equipment costs. Overheads for design, construction, and project management are typically 25% of the direct plant cost. Table 13.1 shows the results of a review of eleven Australian oil and gas processing projects (Brennan & Golonka, 2002). The ratio of standard deviation to average indicates the variability in the data for the plants involved; note that the variability is much smaller for the cases of direct plant cost and total project cost than for individual cost categories.

### 13.4.4  *Estimating Fixed Capital Costs of Plants*

Various approaches to estimating exist, but the *extent of engineering definition* and the *associated design work* in the complete spectrum of plant engineering (including chemical, mechanical, civil, electrical) has a strong influence on the accuracy of the estimate.

Table 13.1 Cost distribution for Australian oil and gas processing projects (Brennan & Golonka, 2002)

| Cost category | Average relative value | Standard deviation | Variability (expressed as standard deviation average) |
|---|---|---|---|
| Total cost of purchased equipment | 1 | – | 0 |
| Equipment installation | 0.19 | 0.06 | 0.32 |
| Piping | 0.60 | 0.19 | 0.31 |
| Instrumentation | 0.23 | 0.11 | 0.46 |
| Electrical | 0.14 | 0.08 | 0.58 |
| Civil | 0.22 | 0.1 | 0.47 |
| Structural steel | 0.14 | 0.09 | 0.61 |
| Buildings | 0.048 | 0.075 | 1.58 |
| Insulation/ fireproofing | 0.089 | 0.066 | 0.74 |
| Painting | 0.014 | 0.011 | 0.78 |
| Direct plant cost | 2.70 | 0.52 | 0.19 |
| Engineering and project management as factor of direct plant cost | 0.25 | 0.065 | 0.26 |
| Total project cost | 3.40 | 0.68 | 0.20 |

**Note:** Values may not sum precisely to totals shown due to rounding.

An early stage of a project is the *feasibility study* stage where a promising project is defined and assessed to determine whether more detailed appraisal is warranted. A later and pivotal stage of the project is the *authorisation* (or *sanction*) stage where the project is assessed by the board of directors (or their delegated authority for a small project) and a decision made whether to commit capital funds and proceed with the project. For the authorisation stage, a level of accuracy required is typically $\pm 10\%$. For this level of accuracy, process flow sheet, mass and energy balances, equipment specifications, piping and instrumentation diagrams, plant layout, and specifications for utilities and effluent treatment would be complete. Much of the civil, electrical, and mechanical design for the plant would also be complete.

For early project estimates, including those for alternative process designs, factored approaches are widely used for both installed equipment and for entire plants, derived from purchased equipment costs. Factors derived from published data (e.g., from Table 13.1) can be used. In selecting values of factors, it is important both to understand the scope of definition of the relevant cost category, and to exercise judgement relevant to the particular plant design.

At individual equipment level or for small sections of plant, cost category factors applied to the purchased cost of equipment depend on equipment size and cost. An example of proposed factors for gas compressors based on an IChemE approach (Gerrard, 2000) is provided in Brennan and Golonka (2002).

## 13.5  Working Capital Costs

While stocks of raw materials and products influence both the fixed capital costs of storage facilities and working capital, the magnitude of working capital for a project is largely independent of fixed capital and should be estimated from first principles. Working capital estimation should be considered in the light of plant operating and business trading conditions, and builds on operating cost estimation rather than fixed capital cost estimation.

Working capital may be broken down into the following items:

- *Raw material stock value*, estimated as the quantity of stock multiplied by its unit value. Raw materials are valued at their purchased cost, inclusive of delivery to site.
- *Value of material stocks held up within the process*. This contribution is normally minor for most chemical processes, but becomes significant where intermediate materials within processes are stored for quality validation.
- *Value of product stocks*, estimated as the quantity of stock multiplied by its cash production cost, summed for each product.
- *Trading balance* between debtors and creditors.
  - ▶ Debtors is another term for accounts receivable. If finished products are paid for at the end of the month following their delivery, the average indebtedness period is 6 weeks. The value of each product is taken at its selling price.
  - ▶ Creditors is another term for accounts payable. If raw materials and utilities are paid for at the end of the month following supply, the average period of credit is 6 weeks, with each raw material or utility valued at their purchased price.

Differing periods for debtors and creditors can occur in business operations with financial implications. Working capital estimation, supplemented by some worked examples, is discussed in more detail in Brennan (1998).

## 13.6  Operating Costs

Once a new or modified plant becomes operational, it will incur costs distinct from those incurred in the investment phase. These costs, termed operating costs, will continue over the duration of the operating life of the plant. It is important to define the *scope of operating costs* (or system boundary applied) which may encompass

- production (gate to gate)
- production inclusive of product distribution
- operation inclusive of all business costs, for example, incorporating
  - ▶ research and development costs
  - ▶ product selling expenses
  - ▶ royalties (payments to technology providers)
  - ▶ corporate administration, or the cost of managing the business.

### 13.6.1  *Simplified Cost Model*

A simple but useful model for gate to gate production cost in a process plant is

$$C = \sum(Rr) + \sum(Eg) + \frac{Mm}{P} + \frac{kI}{P} + [\text{Packaging}] \qquad (13.2)$$

where

$C$ = production cost ($/t product)

$R$ = raw materials consumption (t/t product)

$r$ = unit cost of raw material ($/t raw material)

$E$ = energy consumed (MJ/t product) or alternatively, utility consumption (e.g., t steam/t product)

$g$ = unit cost of energy ($/MJ) or utility (e.g., $/t steam)

$M$ = number of persons employed

$m$ = average cost per employee ($ per person/yr) including payroll overheads

$I$ = fixed capital investment ($)

$k$ = factor to account for annual capital-dependent costs including maintenance, insurance, depreciation

$P$ = annual production rate (tonnes/yr)

Packaging (packages + packaging labour) may be an additional cost for products requiring packaging as distinct from products sent to bulk storage.

**Capacity Utilisation** $U = P/Q$, where $Q$ = annual plant production capacity, is a key determinant of profitability. Profitability is maximised as $U$ approaches a value of 1 (or 100%), that is, where the plant produces at a rate approaching full production capacity. Capacity utilisation may be restricted by

- limitations in market demand
- plant production limitations involving one or more of
  - ▶ raw materials shortage
  - ▶ malfunctioning plant or equipment
  - ▶ plant downtime for maintenance
  - ▶ strikes by operational workers.

## 13.6.2 *Classification of Production Costs*

Production costs may be classified according to

- **variable or fixed** (dependence on production rate)
- **capacity (or scale) dependence**
- **marginal or incremental costs** (incurred per increment of production)
- **cash costs** (with depreciation costs excluded).

*Variable and fixed production costs* are classified according to dependence on production rate. For a given plant, those costs expressed in $/year (as distinct from $/tonne product) which vary with change in production rate are termed *variable* while those costs in $/yr which are unaltered with change in production rate are termed *fixed*. Costs of raw materials and utilities are normally classified as variable. Personnel costs and capital-dependent costs are normally classified as fixed. Note that these are broad approximations which must be more closely examined in detailed cost studies. For example, when some process plants operate below full capacity, some economies in energy and raw materials per tonne of product can be achieved. Again, when plants are shut down, some electricity will be consumed for lighting and for certain essential plant functions, for example, where refrigerated product storage is employed.

## 13.6.3 *Estimation of Production Costs*

Variable costs are estimated based on estimates of raw materials and utilities consumptions and their respective unit costs. Consumptions are based on mass and energy balance and flow sheet considerations. Additional allowances are made at the project evaluation phase for operational inefficiencies and contingencies. In the case of utilities, these additional allowances can amount to as much as 20% of flow-sheet-based estimates.

Fixed capital-dependent costs such as maintenance, insurance and property taxes are estimated as percentages of fixed capital. Mid range estimates are 4% to 6% for maintenance, 1% to 2% for insurance, and 2% for property taxes. More detailed discussion of these costs, costs of personnel, and costs of royalties, research and development, selling expenses, and corporate administration are provided in Brennan (1998) and other process economics texts.

### 13.6.4 *Depreciation*

Depreciation is an allowance for the erosion in value of an asset over its life. For a process plant, the value of the fixed capital investment declines over its life due to physical wear and economic influences. The economic life can be determined by

- the life of a key resource or raw material supply
- product obsolescence or process obsolescence
- lack of competitiveness in scale or cost structure
- excessive maintenance costs
- excessive costs of compliance with tightening environmental legislation.

Depreciation is not an 'out of pocket' expense or a direct component of cash flow, but an allowance made by the company operating a process plant. It can conveniently be calculated as an annual cost using the 'straight line' method, by dividing the fixed capital cost of the plant by the operating life. Thus, for a plant costing $100 million and having an operating life of 10 years, the annual depreciation cost will be $10 million.

Depreciation can be claimed as a tax depreciation allowance when a company is paying corporate tax on its annual profits. Allowable tax depreciation rates are set by government taxation authorities. Note that these rates may differ from those used internally by companies, and can differ from one country to another.

### 13.6.5 *Capital Recovery*

The cost of capital is normally accounted for in cash flow evaluations using net present value (NPV) as an index of profitability.

However, capital recovery can be considered an operating cost and estimated using a capital recovery (CR) factor as a multiplier of capital investment. The CR factor accounts for the *annual* cost of recovering a capital investment over a life of $n$ years where the interest cost of capital (on a real basis and after tax) is $i\%$ per annum. The CR factor $f$ thus accounts for both the interest and principal repayment, and is defined as

$$f = \frac{i(1+i)^n}{(1+i)^n - 1} \tag{13.3}$$

Thus, for an interest cost $i$ of 8%/annum and a life $n$ of 15 yr, $f = 0.117$. The value of $i$ for a project depends on the source of the finance, the extent to which loan and equity funds are used, and the perceived risk of the project.

Note that while accounting for CR costs may be useful in certain cases within operating cost estimates, CR costs should *not* be included in operating costs within cash flows, *where the cash flows are discounted by the time value of money.*

### 13.6.6 *Worked Example: Cost of Utilities Generation*

Equation 13.2 can be used for estimating costs of manufacturing products and also for costs of generating utilities. This example explores the operating cost of cooling water utility.

A recirculated cooling water system comprises a cooling tower, cooling water recirculation pumps, and associated reticulation piping (see Fig. 6.6.). Evaporation, entrainment, and purge losses require a make-up of fresh water of 3.75% of the circulating water flow rate. Given the following estimates of cost and performance, make an approximate estimate of the cost of cooling water per $m^3$ of recirculated cooling water (cw).

**Cost and performance estimates**

| | |
|---|---|
| Flow rate of recirculated water | $0.65 \ m^3/s$ |
| Fixed capital cost of cooling tower, recirculation pumps and piping | $4.0 million |
| Electricity consumption in water recirculation and cooling tower fans | $0.6 \ kWh/m^3$ cw |
| Unit cost of make-up water | 90 cents/$m^3$ |
| Unit cost of electricity | 7 cents/kWh |

**Solution**

| | |
|---|---|
| Cost of make-up water | $0.0375 \times 90 = 3.4$ cents/$m^3$ cw |

| | |
|---|---|
| Electricity cost | $0.6 \times 7 = 4.2$ cents/m$^3$ cw |
| Annual cost of maintenance, insurance, property taxes and depreciation estimated in total at 16% fixed capital | $0.16 \times 4.0 = \$0.64$ million/yr |
| Assuming cooling water circulates continuously for 365 days per year, cost per m$^3$ circulating water | $0.64 \times 10^6 \times 100/0.65$ $\times 3600 \times 24 \times 365$ $= 3.1$ cents/m$^3$ cw |
| **Total cost of recirculated cooling water** | $= 3.4 + 4.2 + 3.1$ $= \mathbf{10.7}$ **cents/m$^3$ cw** |

This assessment excludes costs of operating labour, plant overhead charges, and chemical or effluent treatment, which together are judged to be a minor part of total costs. This simplified approach can be applied to estimating costs of other utilities. The model can also be extended to explore effects of capacity utilisation. If capacity utilisation were limited to 50%, for example, due to seasonal feedstock supply to the process plant, fixed capital-related costs per m$^3$ cooling water would double to 6.2 cents/m$^3$, bringing the total cost to 13.8 cents/m$^3$.

### 13.6.7 *Environmental Management-Related Costs*

Traditional approaches to operating cost estimation account for raw materials, energy, personnel, and capital-related costs. These estimates may exclude some costs associated with waste management or environmental control. The US Environment Protection Agency (US EPA) has developed a classification of costs to encourage a 'total cost assessment' where

Tier 0 — usual process plant costs
Tier 1 — hidden costs of monitoring, paperwork, permit requirements
Tier 2 — liability costs of penalties and fines, future liabilities
Tier 3 — costs derived from consumer responses, employee relations, corporate image.

### 13.6.8 *Cost Sheet Summary*

Table 13.2 is an example of a simplified cost data sheet of estimated production and operating costs. More details can be added as required for different estimates, for example, to cover additional raw material and

Table 13.2   Summary cost sheet for production or operating costs

| Product | | | Annual cost ($ million) | Cost per tonne $/t |
|---|---|---|---|---|
| **Plant capacity** (t/yr) | | | | |
| Fixed capital IBL ($ million) | | | | |
| Fixed capital OBL ($ million) | $ mill | | | |
| **Total fixed capital** ($ million) | $ mill | | | |
| **Production cost** | | | | |
| **Raw materials** | Unit usage (t/t product) | Unit cost ($/t) | | |
| **Total raw materials cost** | | | | |
| Utilities | Unit usage (per t product) | Unit cost ($/unit) | | |
| Steam (t) | | | | |
| Cooling water (m$^3$) | | | | |
| Electricity (MWh) | | | | |
| Effluent treatment | | | | |
| **Total utilities cost** | | | | |
| **Total variable cost** | | | | |
| Process labour | Number | Salary | | |
| Operators per shift | | | | |
| Number of shift teams | | | | |
| Total shift operators | | | | |
| Day operators | | | | |
| **Process labour cost** | | | | |
| **Payroll overheads** | % Wages | | | |
| | % Fixed capital | % Labour | | |
| Maintenance | | | | |
| Plant overheads | | | | |
| Insurance | | | | |
| Property taxes | | | | |
| Book depreciation | | | | |
| **Total fixed cost** | | | | |
| **Total production cost** | | | | |
| Corporate administration | | | | |
| Research, development | | | | |
| Selling expenses | | | | |
| Royalties | | | | |
| **Total operating cost** | | | | |

utility costs, effluent treatment costs, hidden costs associated with wastes, and further details of fixed or non-manufacturing costs. It is important to summarise costs both as *annual costs* and *costs per unit of production*. It is also important to provide a transparent summary of the key unit cost assumptions (for raw materials, utilities, and labour) and factors used for estimating fixed costs or non-manufacturing costs; these factors are commonly applied to fixed capital costs, labour costs, or sales revenue. The operating cost summary sheet is an important working and communication document for any investment concept, proposal, or project.

## 13.7 Revenue or Benefits Estimation

Projects may be driven by any one of several objectives. These include markets for products, opportunities for reduced operating costs, improved safety, or improved environmental performance.

In the case of markets for products, forecasts are needed for market volume, the share of that volume which the project venture is seeking to capture, and the selling price of the product. Sensitivity studies of project profitability have consistently identified product selling price and product sales volume as the two most influential variables in determining profitability.

Demand for a process industry product depends on a complex mix of factors including

- selling price
- product quality
- population and its prosperity as well as consumer behaviour
- extent of industrial development and the capability to further process materials through the manufacturing chain
- competition from alternative products.

Products can be usefully classified according to market volume and product differentiation (Brennan, 1998). Commodities are high-volume products traded globally to the same specification and cover many of the well-known intermediate chemicals like ethylene, caustic soda, and sulphuric acid. Figure 13.1 shows the diffusion of ethylene into various end-use products. Determining the market volume for ethylene, for example, would require estimating the demand for its downstream products. *End-use analysis* is a quantitative method for estimating market volume for a product in which product end uses are identified and projected into the future.

Fine chemicals are undifferentiated but synthesised in rather smaller volumes. Examples of fine chemicals include many pharmaceuticals and food additive products. Specialty chemicals are low volume products but

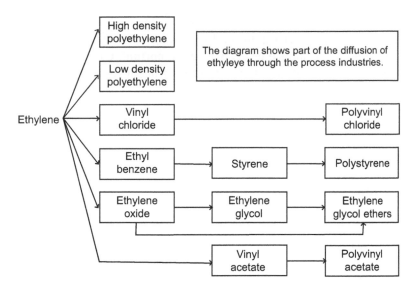

**Figure 13.1** Common end uses for ethylene (Brennan, 1998). Reprinted with permission of IChemE.

differentiated and developed for particular applications; examples include catalysts and chemicals for boiler feedwater and cooling water treatment.

Selling price projections for products requires consideration of many factors including whether the product is new or established, the extent of competition, production cost projections, and potential for technological change. These factors and useful data sources for price and market volume projections are discussed in Brennan (1998).

In the case of investing for reduction in operating cost, careful consideration of current and projected operating costs is required including consultation with operating and maintenance staff. Any new investment involving added equipment will necessarily involve incremental maintenance and other fixed capital-related costs.

In the case of safety and environmental projects, the operating costs resulting from the 'no investment' option must be considered as well as those for the proposed investment, as part of the project evaluation.

## 13.8 Engineering for a Movable Target

An additional complication arises in environmentally driven projects from the horizon of progressively tighter emission limits. Is it better to just meet *existing* limits, or to meet *anticipated future* limits while incurring extra cost? Here, the marginal capital and operating costs of achieving better environmental performance levels become important. Cost implications of mandatory future investment to meet more stringent regulations must also be considered. Other benefits such as operating cost reduction may accrue from investment and should be carefully evaluated. The case of meeting regulatory sulphur levels in transport fuels is an example of engineering for a moving target.

## 13.9  Profitability

Profitability is commonly assessed as return on investment, or annual profit divided by capital investment expressed as a percentage per annum. For *detailed evaluations* required of *major investments*, profitability is based on cash flow projections over the life of the investment and accounts for the *time value of money*.

**What do we mean by cash flow?**
Cash flow is the net flow of money into or out of a project (or a company) over a given time period. The commonly used time period in economic evaluation of projects is one year. Cash outflows (e.g., capital expenditure on plant, expenditure on increased raw material stocks, tax payments) are considered negative, while cash inflows (e.g., income after tax, recovery of working capital at the end of the project life) are considered positive.

**Depreciation** is not a cash flow in itself or a component of cash flow as discussed under 13.6.4. A tax depreciation allowance, however, is a permissible deduction from income for tax assessment in most countries, and as such influences the magnitude of tax payments, and hence of after tax cash flows.

**What do we mean by the time value of money?**
Money has interest bearing capability. A fund of $100 today becomes $105 in 1 year's time if the real interest rate is 5% per annum. Conversely, an incoming payment of $105 occurring in 1 year's time has a present value of $100.

The present value (PV) of a future cash flow $C_t$ occurring in $t$ years time where the cost of money is $i$ per annum is estimated as

$$PV = \frac{C_t}{(1 + i/100)^t}$$

The NPV for the project is the sum of individual present values of all cash flows summed over the life ($k$ years) of the project.

Thus,

$$\text{NPV} = \sum_{t=0}^{t=k} \frac{C_t}{(1 + i/100)^t}$$

A positive NPV implies that a net cash benefit is obtained as a result of the project; hence the project is profitable. The value of $i$ for which the NPV is zero is called the discounted cash flow rate of return (DCFR) or alternatively the internal rate of return (IRR).

Profitability measures derived from cash flow projections are commonly expressed in terms of

- net cash generated, accounting for the time value of money (NPV)
- rate of return (DCFR or IRR)
- payback time (time to recover capital investment).

Details of profitability assessment based on the time value of money are provided in Brennan (1998) and other process economics texts. A simple measure of profitability is

$$\text{return on investment (ROI)} = \frac{\text{annual profit}}{\text{total capital investment}}$$

where annual profit = sales revenue generated minus operating costs.

ROI is normally expressed as %/yr. Annual profit might be based on cash flow, or may be based on operating costs which include depreciation. ROI may be before or after tax. It is important to *make these distinctions* both in *specifying an estimated ROI* and also *interpreting ROI estimates made by others*.

*Profitability* is fundamental for business survival and exerts strong influence on investment projects and related planning decisions (e.g., choice of product, technology, manufacturing site). Profitability considerations can constrain initiatives in waste treatment, cleaner production, and recycling projects such as

- treating $SO_2$ emissions from sulphide ore smelters
- substitution of mercury cell chlorine plants with membrane cell plants
- recycling of plastics, where costs and quality of recycled polymer must be compared with costs and quality of virgin polymer
- converting waste into usable product.

## 13.10 Case Example

This case example is provided to illustrate the estimation and documentation of operating costs, the estimation of working capital, the development of a cash flow table, and profitability assessment based on the estimated cash flows.

**Problem.** A membrane cell chlor alkali plant of 300 tonnes chlorine per day capacity has been proposed as part of a chemical industry redevelopment scheme. For every tonne of chlorine produced, 1.12 tonnes of caustic soda is also produced. The plant is to be operated continuously. The following performance and costs have been estimated, the costs being expressed in Australian $.

**Performance and Cost Data**

| | | | |
|---|---|---|---|
| **Plant capacity** | 300 tonnes chlorine per day | | |
| **Fixed capital** (spent uniformly over 2 years) | | | |
| Inside battery limits | $130 million | | |
| Outside battery limits | $40 million | | |
| **Raw materials** | Consumption | Unit cost | **$/t chlorine** |
| Salt | 1.7 t/t chlorine | $30/t | |
| Other chemicals | | | 11 |
| **Utilities** | Consumption | Unit cost | |
| Electricity | 3.0 MWh/t chlorine | $70/MWh | |
| Other utilities | | | 4 |
| **Membranes replacement** | | | 20 |

| **Process labour** | | Annual wage per operator |
|---|---|---|
| Shift labour | 4 operators per shift | $70,000 |
| Day labour | 8 operators/day | $50,000 |
| Payroll overheads | 40% of wages | |
| **Plant overheads** | 80% of process labour costs (inclusive of payroll overheads) | |
| **Annual cost of maintenance, insurance, property taxes** | 6.5% total fixed capital | |
| **Annual cost of corporate administration, selling expenses, research, and development** | 1.5% of total fixed capital | |

The plant is to be constructed in 2012 and 2013 to commence production in January 2014. Forecast annual sales volume for chlorine are 70,000 tonnes in 2014 and 100,000 tonnes in 2015 and subsequent years until the end of project life, assumed for December 2028. Anticipated selling price of caustic soda is $450/tonne, and of chlorine is $430/tonne.

An electrochemical unit (ECU) comprises 1 tonne of chlorine and 1.12 tonne of caustic soda. On the basis of the cost and performance data provided, estimate

(a) the cash operating cost per year, and per ECU
(b) the working capital requirements of the plant at 100,000 tonnes chlorine per year output.

Assume for working capital requirements:

- salt stock equivalent to 1 month production
- caustic soda stock equivalent to 1 month production
- negligible chlorine stock and work in progress stock
- 6 weeks debtors; 1 month creditors.

On the basis of the data provided, estimate the cash flows after tax over the anticipated life of the project. Assume that inflation will be negligible, the corporate tax rate is 30% of taxable income, and that tax depreciation allowance is straight line at 6.7% per annum of fixed capital until the plant is fully depreciated. Assume that no investment allowance applies and that tax is paid in the same year in which the related income is generated. On the basis of the estimated cash flows, and assuming the cost of capital for the project is 10% per annum, make an assessment of the project's perceived profitability. (Note that tax allowances and rates differ from one country to another, and can also change over time with change in government policy for individual countries.)

## Solution
### Sales revenue estimate

$$\text{Revenue from sale of 1 ECU} = 450 \times 1.12 + 430 = \$934$$

$$\text{Revenue at full plant output} = 100,000 \times (1 \times 430 + 1.12 \times 450)$$
$$= \$93.4 \, \text{million}$$

### Operating cost estimate
Operating costs are based on the data provided in the problem statement and are summarised in Table 13.3.

### Working capital estimate
Working capital is estimated from assumptions stated in problem and derived from operating cost estimate

|  | Period | Cost ($ million) | Cost ($ million) |
|---|---|---|---|
| Salt stock | 1 month | $1/12 \times 5.1$ | 0.43 |
| Caustic soda stock | 1 month | $1/12 \times 47.03 \times 1.12/2.12$ | 2.07 |
| Debtors | 6 weeks | $6/52 \times 93.4$ | 10.78 |
| Creditors | 1 month | $29.6/12$ | −2.47 |
| **Total** | | | **10.81** |

The cash flow table (Table 13.4) summarises the fixed capital investment, working capital investment, sales revenue, and operating costs over the life of the project. From these estimates, the cash flow before tax is calculated. The tax depreciation allowance during operational years is then claimed, the tax payment calculated, and the resulting after tax cash flow calculated. The cumulative cash flow after tax indicates the net cash position for the project at various stages over the life of the project. Inflation has been ignored for the purpose of simplification. Accounting for inflation in cash flow projections is discussed in Brennan (1998).

Cash flow tables are conveniently constructed on spreadsheets. Financial functions in the spreadsheet software enable the calculation of NPV or DCFR (also termed the IRR). The NPV and DCFR calculations can be explored for

Table 13.3   Operating cost summary sheet for chlor-alkali plant

| Product | Chlorine | Caustic soda | | |
|---|---|---|---|---|
| **Plant capacity** | t/yr | 100,000 | | |
| Fixed capital IBL | $mill | 130 | | |
| Fixed capital OBL | $mill | 40 | | |
| **Total fixed capital** | $mill | 170 | | |
| Production cost | | | **Annual cost ($ million)** | **Cost per ECU ($/ECU)** |
| **Raw materials** | Unit usage t/t product | Unit cost $/t | | |
| Salt | 1.7 | 30 | | 51 |
| Chemicals | | | | 11 |
| **Total raw materials cost** | | | | 62 |
| **Utilities** | Unit usage Per t product | Unit cost $/unit | | |
| Electricity (MWh) | 3 | 70 | | 210 |
| Other | | | | 4 |
| **Total utilities cost** | | | | 214 |
| **Membrane replacement** | | | | 20 |
| **Total variable cost** | | | 29.6 | 296 |
| | | Annual | | |
| **Process labour** | Number | Salary $/yr | | |
| Operators per shift | 4 | | | |
| Number of shift teams | 4 | | | |
| Total shift operators | 16 | 70,000 | 1.12 | |
| Day operators | 8 | 50,000 | 0.40 | |
| **Process labour cost** | | | 1.52 | |
| **Payroll overheads** | % wages | 40 | 0.61 | |
| | % fixed cap | % Labour | | |
| **Maintenance plus** | | | | |
| Insurance plus | | | | |
| Property TAXES | 6.5 | | 11.05 | |
| **Plant Overheads** | | 80 | 1.70 | |
| **Total fixed cost** | | | 14.88 | |
| **Total production cost** | | | 44.48 | |
| **Non-manufacturing** | % fixed cap | | | |
| Corporate administration. | | | | |
| Research & development and | 1.5 | | 2.55 | |
| selling expenses | | | | |
| **Total operating cost** | | | 47.03 | 470.30 |

different project life scenarios, and the NPV calculation for different discount rate scenarios as desired.

The calculated cash flows in Table 13.4 indicate that the cumulative cash flow after tax becomes positive in the sixth year of operation. The

Table 13.4    Cash flow table for case example

| Year | | 2012 | 2013 | 2014 | 2015 | 2016 | 2017 | 2018 | 2019 | 2020 | 2021 | .... | 2028 |
|---|---|---|---|---|---|---|---|---|---|---|---|---|---|
| Fixed capital | | −85 | −85 | | | | | | | | | | |
| Working capital | | | | −10.81 | | | | | | | | | 10.81 |
| Sales volume ECU/yr | | | | 70,000 | 100,000 | 100,000 | 100,000 | 100,000 | 100,000 | 100,000 | 100,000 | | 100,000 |
| Selling price $/ECU | 934 | | | | | | | | | | | | |
| Sales revenue | | | | 65.38 | 93.4 | 93.4 | 93.4 | 93.4 | 93.4 | 93.4 | 93.4 | | 93.4 |
| Variable costs | | | | 20.72 | 29.6 | 29.6 | 29.6 | 29.6 | 29.6 | 29.6 | 29.6 | | 29.6 |
| Fixed costs | | | | 17.43 | 17.43 | 17.43 | 17.43 | 17.43 | 17.43 | 17.43 | 17.43 | | 17.43 |
| **Cash flow before tax** | | | | **27.23** | **46.37** | **46.37** | **46.37** | **46.37** | **46.37** | **46.37** | **46.37** | .... | **46.37** |
| Tax depreciation rate % | 6.7 | | | | | | | | | | | | |
| Tax dep'n allowance | | | | 11.39 | 11.39 | 11.39 | 11.39 | 11.39 | 11.39 | 11.39 | 11.39 | | 11.39 |
| Taxable income | | | | 15.84 | 34.98 | 34.98 | 34.98 | 34.98 | 34.98 | 34.98 | 34.98 | | 34.98 |
| Tax rate % | 30 | | | | | | | | | | | | |
| Tax payment | | | | 4.75 | 10.49 | 10.49 | 10.49 | 10.49 | 10.49 | 10.49 | 10.49 | | 10.49 |
| **Cash flow after tax** | | −85 | −85 | **11.67** | **35.88** | **35.88** | **35.88** | **35.88** | **35.88** | **35.88** | **35.88** | .... | **46.69** |
| **Cumulative cash flow** | | −85 | −170 | **−158.33** | **−122.45** | **−86.57** | **−50.69** | **−14.81** | **21.07** | **56.95** | **92.83** | .... | **354.80** |
| Discount factor | | **0.909** | **0.826** | **0.751** | **0.683** | **0.621** | **0.565** | **0.513** | **0.467** | **0.424** | **0.386** | | 0.198 |
| Discounted cash flow after tax | | −77.27 | −70.24 | 8.77 | 24.50 | 22.28 | 20.25 | 18.41 | 16.74 | 15.22 | 13.83 | | 9.24 |
| **Cumulative discounted cash flow after tax** | | **−77.27** | **−147.52** | **−138.75** | **−114.25** | **−91.97** | **−71.71** | **−53.30** | **−36.56** | **−21.35** | **−7.51** | .... | **61.97** |

**Notes:** All sums, cash flows are in $ million. Working capital assumed fully spent in year 2014 at start of production year and reclaimed in 2028 in final production year.

Table 13.5  Dependence of net present value on discount rate for case example

| Discount rate (%/yr) | 0 | 6 | 8 | 10 | 12 | 14 | 16 | 18 | 20 |
|---|---|---|---|---|---|---|---|---|---|
| Net present value ($ million) | 354.80 | 137.99 | 95.43 | 61.97 | 35.50 | 14.43 | −2.42 | −15.96 | −26.89 |
| Discounted cash flow rate of return (%/yr) | 15.7 | 15.7 | 15.7 | 15.7 | 15.7 | 15.7 | 15.7 | 15.7 | 15.7 |

cumulative discounted cash flow (or NPV) at the discount rate of 10% per annum becomes positive in the ninth year of operation. While the project is estimated to have a positive NPV, the duration of the payback time (both in undiscounted and discounted terms) indicates that the project is only marginally profitable. The dependence of NPV on discount rate is shown in Table 13.5. The DCFR is 15.7%.

## 13.11  Economies of Scale

Scale has an important influence on both capital and operating costs, for both plant and equipment.

For a plant, economies of scale exist when the capital investment per unit of production capacity decreases with increasing production capacity. Scale economies also exist when operating costs per tonne of product decrease with increased production capacity. These scale economies encourage investment into larger capacities at both equipment and plant levels. Larger capacities of equipment can be achieved through physical size increase and through technology improvement. Increase in capacity for a given physical size is often referred to as intensification and is discussed further in Chapter 15.

A convenient basis for examining the effects of scale on production costs is to consider production costs as the sum of raw materials, utilities, personnel, and capital-dependent costs as indicated in Equation 13.2. Raw materials consumption and utilities consumption per tonne of product are generally independent of capacity; however, personnel employed per tonne of product ($M/QU$) and capital-dependent costs per tonne of product ($kI/QU$) both decrease with increased plant capacity for most cases.

The number of persons employed in a plant increases marginally with larger plant capacity. The dependence of labour on plant capacity will be larger for parallel stream plants than single stream plants reflecting the greater numbers of process and maintenance labour required for a greater number of equipment items. Capital-dependent costs tend to follow a similar dependence on capacity to that of capital investment.

For a set of plant capacity options for a particular technology where the unit costs of raw materials, utilities, and personnel are assumed scale invariant, variable costs (raw materials and utilities) per tonne of product are essentially scale invariant while fixed costs (personnel and capital-dependent costs) per tonne of product are scale dependent.

The dependence of production cost on capacity is then governed by

(a) the dependence of $M$ and $I$ on $Q$
(b) the cost structure, or the dominance of personnel and capital-dependent costs in production cost.

Thus, we may have relatively steep or flat scale curves for production costs for plants of different technologies and products reflecting different cost structures. The effect of scale on production costs with examples of specific technologies has been discussed in detail in the engineering and economics literature, for example, by Brennan (1992).

Problem 5 at the conclusion of Part C explores the effects of scale on electricity generation costs.

A case example for a chemical plant is now explored.

### 13.11.1  *Case Example in Scale Economies*

We explore the capital and operating costs for the chlorine plant case (Section 13.10) previously explored but in this case having a capacity of 50,000 tonne/yr.

$$\text{Fixed capital investment for the plant } I = kQ^b$$

Assume $b = 0.7$ reflecting some mixed parallel stream (electrochemical cells) and single stream (feed purification and product purification) plant configuration.

***Fixed capital IBL*** $= 130 \times 0.5^{0.7} = \$80.0$ million.
Assume fixed capital OBL is in the same proportion to IBL capital as for the larger scale case. Then *fixed capital OBL* $= (40/130) \times 80 = \$24.6$ million.

***Total fixed capital*** $= \$104.6$ million, showing a 20% increase in fixed capital investment per unit of production capacity compared with the larger plant.

Considering operating costs, *variable production costs* can be assumed to be the same as for the larger plant, that is, $\$296/\text{ECU}$.

***Process labour*** requirement is estimated to be proportional to $Q^{0.5}$ reflecting a degree of parallel streaming. Corresponding cost becomes $2.13 \times 0.5^{0.5}$ = \$1.51 million/yr inclusive of payroll overheads, equivalent to \$30.2/ECU.

***Plant overheads*** are estimated to achieve only minor savings for the smaller plant, say $0.9 \times 1.70$ = \$1.53 million/yr equivalent to \$30.6/ECU

***Capital-dependent costs*** are again estimated as 6.5% total fixed capital = $0.065 \times 104.6$ = \$6.80 million/yr, equivalent to \$136/ECU

***Non-manufacturing costs*** are again estimated as 1.5 % fixed capital = $0.015 \times 104.6$ = \$1.57 million/yr, equivalent to \$31.4/ECU

Thus, total fixed costs are estimated as $30.2 + 30.6 + 136 + 31.4$ = \$228/ECU

***Total cash operating cost*** = \$296 + \$228 = \$524.2/ECU showing an 11% increase compared with the larger plant.

Note that the proportion of variable costs in operating costs has decreased from 63% in the larger plant to 56% in the smaller plant. If we were to consider a smaller plant again, say of 25,000 tonnes/year capacity, the proportion of variable costs in total operating costs would decrease even further. Thus, there are shifts in operating cost structure with change in scale. In larger capacity plants, especially for technologies where variable costs are dominant, additional capital is often spent to improve raw materials and energy efficiency to reduce variable costs, even at the expense of increasing fixed operating costs. Thus, larger scale plants do not always mirror smaller scale plants in terms of design detail.

It is also important to note that operating cost penalties for smaller plants may be offset by potential benefits in unit costs of raw materials, utilities and labour which may apply at a given location. Thus, apart from the consideration of matching plant size to market opportunities, there are cases where smaller plants can still be cost competitive with larger plants.

## 13.12 Environmental Externalities

The costs of environmental damage are borne by the community at large. Because these costs fall outside the accounting framework of the 'polluter', they are often called 'external costs' or 'externalities'. Examples include costs of damage caused by

- air pollution
  - ▶ degradation of architectural buildings, structures, landscapes

▶ deterioration of human health, for example, due to respiratory problems
- water pollution
  ▶ loss of marine life and hence food
  ▶ loss of recreational value of a river or lake
- soil pollution
  ▶ costs of eventual remediation
  ▶ loss of biodiversity.

An important current example is the cost to the wider community of global warming and associated climate change due to greenhouse gas emissions. Many of the conditions attributed to climate change such as rising sea levels and extremes of weather patterns including storms and drought, have major 'external cost' implications. These costs fall into a number of categories including preventive costs such as costs of infrastructure to withstand or adapt to extreme weather patterns, loss of land suitable for housing near coastlines, increased insurance premiums, repairs to damaged property, and reduced productivity of agricultural land.

### 13.12.1 *Estimating External Environmental Costs*

Various proposals have been made for estimating external environmental costs. *Contingent valuation techniques*, for example, try to assess people's willingness to pay for environmental amenities using a survey technique. While there are difficulties in estimating external costs due to pollution, it is clear that these costs can be major and are not directly borne by those who inflict them.

The cost of damage caused by pollution from power stations is an example of external costs of electricity generation. A major concern is climate change, but impacts on human health and crop damage are also important. Some estimates of marginal damage costs for emissions varies considerably with European Environment Agency (2007) reporting a range of 19 to 80 EUR/t $CO_2$ for $CO_2$ emissions. Some unit damage costs reported by Keoleian and Spitzley (2006) and drawn from earlier publications for utility generation in mid-western United States are shown in Table 13.5. Keoleian and Spitzley stress that a wide range of values are reported for such unit damage costs.

*Landfill disposal* is another example of an activity involving externalities. Leakage of landfill gas from old landfill sites has in some cases affected nearby residents, causing temporary evacuation from homes. Xu (1998) has estimated external costs of landfill (Table 13.6).

Table 13.6    Estimates of unit damage costs for common electricity utility emissions (Keoleian & Spitzley, 2006)

| Pollutant | Unit damage cost (US$) |
|-----------|------------------------|
| $CO_2$ | $30/t carbon |
| Lead | $1965/t |
| Nitrogen oxides | $218/t |
| Particulates | $2624/t |
| Sulphur oxides | $84/t |

Table 13.7    Estimates of external costs for landfill (Xu, 1998)

| Cost element | Total NPV $ million | Annual equivalent Cost $/year | Unit cost $/t waste |
|--------------|---------------------|-------------------------------|---------------------|
| Waste haulage | | | |
|   a. air pollution | 0.10 | 8,500 | 0.09 |
|   b. traffic accident | 0.19 | 15,700 | 0.17 |
| Site disamenity | 4.54 | 380,000 | 4.22 |
| Risk of leachate | 0.52 | 43,400 | 0.48 |
| LFG emissions | | | |
|   a. $CO_2$ | 1.24 | 104,300 | 1.16 |
|   b. $CH_4$ | 3.05 | 255,100 | 2.83 |
| **Total cost** | **9.64** | **807,000** | **8.95** |

**Note:** Costs are in Australian $ 1998.

## 13.12.2  *Emission Taxes and Emission Trading Schemes*

Apart from measures to limit emissions by regulatory control, much interest has focused recently on the potential for use of emission taxes and emission trading schemes. This is particularly so in the context of reducing greenhouse gas emissions. Emission charges and trading schemes are discussed further in the context of economic instruments in Chapter 15 under government legislation. Problem 5 at the conclusion of Part C explores the effect of a carbon tax on electricity generation costs for power stations using different fuels, power generation cycles, and scales.

## 13.13  Life Cycle Costs of Projects

Life cycle costs account for the full cost of a project over its entire life. For a process plant this includes costs incurred in the planning and design phase, plant construction, the operating life of the plant, and the decommissioning and dismantling of the plant (including any site rehabilitation). The operating phase of the plant will include operation, maintenance, and any modifications to improve capacity or performance. Decommissioning will include dismantling and disposal of plant and equipment, and any site remediation required.

Estimation of capital investment and operating costs have been discussed in this chapter. Costs of planning and design are increased where unforeseen delays and difficulties are encountered. Decommissioning costs depend on the technology and site and can be extensive in some cases, for example, large mining projects, waste disposal sites, and nuclear power stations.

## 13.14 Conclusions

Economic evaluation encompasses fixed capital and working capital estimation, operating cost estimates, market evaluation and its implications for revenue estimation, and profitability assessment. The key principles underpinning such estimates have been outlined and discussed. These various estimates draw on design, operational, and commercial expertise and are typically generated by different personnel with distinct skill and experience profiles.

Economic evaluation over the life of a future project involves projections and estimates into the future with attendant risks. These aspects are discussed further in Chapter 16 in the context of the development and assessment of projects.

## References

Brennan, D. J. (1992) Evaluating scale economies in process plants, *Trans. IChemE.*, **70**(Part A), 516–526.

Brennan, D. (1998) *Process Industry Economics*, IChemE, Rugby, England.

Brennan, D. J. and Golonka, K. A. (2002) New factors for capital cost estimation in evolving process designs, *Trans.IChemE.*, **80**(Part A), 579–586.

European Environment Agency (2007) *EN35 External costs of electricity production*, http://www.eea.europa.eu/data-and-maps/indicators/en35.

Gerrard, A. M. (2000) *Guide to Capital Cost Estimating*, IChemE, Rugby, England.

Keoleian, G. A. and Spitzley, D. V. (2006) Life cycle based sustainability metrics (Chapter 7), in *Sustainability Science and Engineering. Defining Principles* (ed. Abraham, M. A.), Elsevier.

Peters, M. S., Timmerhaus K. D., and West, R. E. (2003) *Plant Design and Economics*, McGraw-Hill, New York.

Sinnott, R. and Towler, G. (2009) Costing and project evaluation (Chapter 6), in *Chemical Engineering Design*, Elsevier, Oxford, England.

Xu, X. (1998) *Economic Analysis of Landfills*, PhD thesis, University of Queensland.

Appendix 1.   Estimates of plant cost inflation indices for Australia 1982–2008

| Financial year June | Materials price index | Producer price index | Average weekly earnings | Plant cost index | Consumer price index |
|---|---|---|---|---|---|
| 1981 | —— | approx | —— | —— | 100.0 |
| 1982 | 125.4 | | 283.75 | 100.0 | 110.4 |
| 1983 | 139.6 | | 324.15 | 112.8 | 123.1 |
| 1984 | 147.3 | | 349.45 | 120.3 | 131.6 |
| 1985 | 155.5 | | 376.08 | 128.2 | 137.2 |
| 1986 | 167.9 | | 399.48 | 137.3 | 148.7 |
| 1987 | 180.9 | | 427.98 | 147.5 | 162.6 |
| 1988 | 196.8 | | 454.48 | 158.6 | 174.5 |
| 1989 | 214.9 | | 487.30 | 171.6 | 187.3 |
| 1990 | 231.9 | | 520.95 | 184.3 | 202.3 |
| 1991 | 243.7 | | 555.40 | 195.1 | 213.0 |
| 1992 | 245.2 | | 580.75 | 200.2 | 217.0 |
| 1993 | 245.7 | | 591.08 | 202.0 | 219.2 |
| 1994 | 249.2 | | 608.78 | 206.5 | 223.5 |
| 1995 | 255.9 | | 634.00 | 213.6 | 230.7 |
| 1996 | 261.2 | | 662.43 | 220.6 | 232.9 |
| 1997 | 262.4 | | 688.23 | 225.4 | 236.0 |
| 1998 | 264.7 | | 716.65 | 231.0 | 236.0 |
| 1999 | 266.9 | 100.0 | 743.68 | 236.3 | 238.9 |
| 2000 | 269.2 | | 768.55 | 241.3 | 244.6 |
| 2001 | 269.9 | | 808.57 | 247.9 | 259.3 |
| 2002 | 275.0 | | 853.57 | 257.1 | 266.8 |
| 2003 | 286.4 | | 897.40 | 269.0 | 275.0 |
| 2004 | 289 | | 952.83 | 278.7 | 281.4 |
| 2005 | 300.7 | 116 | 1008.10 | 292.4 | 288.2 |
| 2006 | | 120 | 1032.00 | 300.9 | 299.2 |
| 2007 | | 124 | 1091.03 | 314.5 | 304.4 |
| 2008 | | 131 | 1131.40 | 329.2 | 317.4 |
| 2009 | | 129 | 1201.10 | 336.8 | 312.5 |
| 2010 | | 132.5 | 1256.70 | 349.1 | 321.0 |

**Notes:** The Australian plant cost index has been estimated based on a 50% weighting of the price index for materials used in building other than house building, and a 50% weighting of average weekly earnings (ordinary-time earnings for full- time adults). The materials price index and average weekly earnings have been averaged over the stipulated time periods. The Australian Consumer Price Index for the period 1981 to 2008 is included for comparison. ABS has ceased publishing the materials price index for buildings other than houses but published a producer price index (similar data) from 2000.

# Chapter 14

# Sustainability Assessment

## 14.1 Introduction

Sustainability assessment involves many perspectives but three key steps can be identified. Firstly, there is the *integration* of economic, environmental, and social assessments and the assurance that adequate levels of performance in all three criteria are met. Secondly, there is the *projection* into the future to ensure that standards in economic, environmental, and social criteria do not deteriorate over future time horizons; this involves postulating a number of scenarios against which proposed projects or plans (or lack of them) can be examined. Thirdly, there is the establishment of sustainability indicators which enable *quantification* of performance and progress over time.

## 14.2 Common Threads in Enviro-Economic Assessment

*Enviro-economic assessment*, or the integration of environmental and economic assessments, is an important aspect of the overall integration process. It is important to note some important common threads in environmental and economic assessment. For chemical processes, *raw materials and utilities consumptions* per tonne of product provide *key inputs* for *both environmental and economic assessment*. In *economic assessment*, consumptions are multiplied by unit costs of raw materials and utilities to determine *variable costs* for the process, as outlined in Chapter 13.

*Sustainable Process Engineering: Concepts, Strategies, Evaluation, and Implementation*
David Brennan
Copyright © 2013 Pan Stanford Publishing Pte. Ltd.
ISBN 978-981-4316-78-1 (Hardcover), 978-981-4364-22-5 (eBook)
www.panstanford.com

In *environmental assessment,* utilities consumptions are further defined in terms of inputs, mainly fuel and water, to become part of the life cycle inventory, as outlined in Chapters 10 and 11. Process raw materials make up the other part of the life cycle inventory for inputs. In environmental assessment, *emissions from utilities generation* and *process effluents* make up the *outputs* of *the life cycle inventory.* Life cycle inventory imply resource depletion impacts for inputs, and a spectrum of damage impacts for outputs. In certain cases, there may be *cost allocations* for process and utility emissions, either as *effluent treatment costs* or *taxes* on certain emissions, for example, carbon dioxide.

Figure 14.1 shows inputs and outputs for a process where a main feedstock is upgraded into a saleable product. The inputs and outputs have both economic and environmental consequences. The specific case of diesel hydrotreating was considered in Chapter 11 as a case example in life cycle assessment. Figure 11.5 showed life cycle inventory data for the hydrotreating process, where diesel is upgraded by treatment with hydrogen to remove sulphur. The inventory data includes input and output data for both process and utility streams, providing a basis for both environmental impact assessment and operating cost estimation. Consumptions of hydrogen and utilities enable estimation of costs for these inputs; if applicable, carbon dioxide taxes or emissions treatment costs could also be applied to the defined emissions.

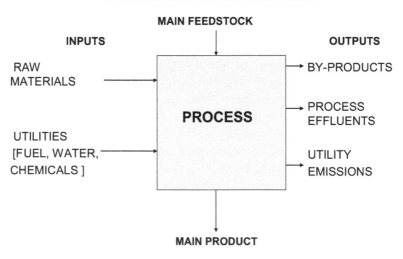

**Figure 14.1** Economic and environmental inputs and outputs for a process.

## 14.3 Cost Benefit Approach to Enviro-Economic Assessment

Traditionally, economic assessment of a process industry project has been over project life on a cash flow basis, accounting for capital expenditure, sales revenue, operating costs, taxation, and the cost of capital. This approach has been outlined in Chapter 13. In public expenditure projects, cost benefit analysis has been more widely used to assess benefits to society in conjunction with costs. Cost benefit analysis can be usefully applied to environmentally driven initiatives such as cleaner transport fuels, where benefits in human and environmental health are achieved but with increased costs in petroleum refining. Cost benefit analysis also has a useful role in enviro-economic assessment when making process engineering decisions. Such decisions may be directed towards improving a particular process under study, towards selecting alternative processes, or evaluating a new process or product initiative.

When evaluating process engineering decisions, trade-offs can occur in economic criteria between capital and operating costs, in environmental criteria between one environmental impact and another, and across criteria between costs and environmental performance. It is important to identify and if possible quantify benefits and burdens, whether economic or environmental, to aid decision making and resolve trade-offs. A simple checklist for identification of benefits, burdens and trade-offs is shown in Table 14.1. For both economic and environmental assessment, system definition incorporates the whole product life cycle. However, the cost benefit approach is applicable to any segment of the product life cycle, including an individual process plant.

Table 14.1 Checklist for identification of benefits, burdens, and trade-offs

| Case X |
| --- |
| **Definition and brief explanation of design option** |
| **Environmental benefits** |
| **Environmental burdens** |
| **Economic benefits** |
| **Economic burdens** |
| **Trade-offs** |
|     Economic |
|     Environmental |
|     Environmental vs economic |

## 14.4 Quantifying Benefits and Burdens

Economic benefits and burdens can be quantified as sales revenues, capital costs, operating costs, cash flows and net present values. Environmental benefits and burdens can be quantified as environmental impact potentials using life cycle assessment methodology; inventory data can be used as the basis for providing raw or normalised scores in relevant impact categories. The sum of normalised scores can also be used in conjunction with individual impact scores as an approximate indicator of overall impact potential.

Economic and environmental measures can be left in their respective units. In some studies, monetary values have been assigned to emissions or impact categories, enabling enviro-economic assessment within a financial model. Cases include power generation industries, for example, within the Europe-based ExternE project as reported in Chapter 13. Such values are inherently approximate and often subjective, but highlight an increased recognition of the need to identify and internalise external costs for incorporation into decision making. While monetary values have been postulated for impacts resulting from emissions, published estimates for resource depletion impacts are rare.

## 14.5 Case Examples in Enviro-Economic Assessment

We now examine some process industry cases where economic and environmental benefits and burdens have been identified and quantified.

### 14.5.1 *Case 1: Sulphuric Acid Manufacture*

This case involves a single process plant (micro scale in industrial ecology terminology). The process has been discussed in Chapter 4 in conjunction with a simplified flow sheet based on sulphur feedstock, which is oxidised completely to sulphur dioxide. Subsequent conversion of sulphur dioxide to sulphuric acid depends on the reversible oxidation to sulphur trioxide

$$SO_2 + 0.5O_2 \longleftrightarrow SO_3 \quad \Delta H = -98.5 \, \text{kJ/mol}$$

followed by the absorption of sulphur trioxide in water (within a 98%w/w sulphuric acid solution)

$$SO_3 + H_2O \rightarrow H_2SO_4 \quad \Delta H = -130.4 \, \text{kJ/mol}$$

Exhaust gases from the final $SO_3$ absorption column contain $SO_2$, resulting from the reversible nature of the $SO_2$ oxidation reaction.

In the evolution of sulphuric acid manufacturing technology, $SO_2$ emissions have been minimised using two basic strategies:

1. Increasing the number of reaction stages.
2. Providing an intermediate stage of $SO_3$ absorption to push the equilibrium reaction to the right in later reaction stages. Absorption occurs at a lower temperature than reaction, requiring heat exchange between the reactor and intermediate absorption column.

The first strategy requires incremental capital for additional reaction stages and additional gas cooling between reaction stages, and incurs additional energy to overcome the increased gas pressure drop through the added equipment.

The second strategy requires incremental capital for an additional absorption column and associated heat exchangers, and incurs additional energy consumption due to increased gas pressure drop.

In both strategies, operating costs increase due to increased energy consumption and increased maintenance derived from added equipment.

## Using benefit and burden analysis

### Environmental benefits of the two strategies

- Reduced $SO_2$ emissions to atmosphere implying reductions in acidification potential and toxicity

### Environmental burdens of the strategies

- Increased energy consumption due to increased pressure drop through additional reactor stages, absorption column, heat exchange equipment, and piping. This implies increase in global warming potential and resource depletion where energy is derived from fossil fuel.

### Economic benefits

- Small increase in acid production and related sales revenue

### Economic burdens

- Increased capital cost through provision of increased equipment and piping
- Increased operating cost through increased utilities cost, maintenance cost, and other capital dependent costs

**Trade-offs**

*Economic*: Increased sales revenue versus increased capital and operating costs

*Environmental*: Reduced acidification and toxicity versus increased global warming and resource depletion

*Enviro versus economic*: Reduction in net environmental impact versus increased capital and operating costs.

Figure 14.2 shows the relationship between operating costs and environmental impact for a range of design options in the manufacture of sulphuric acid from metallurgical smelter gases (Golonka & Brennan, 1997). An environmental index has been calculated as the sum of normalised impacts for acidification and global warming determined using life cycle assessment methodology. Normalisation has been carried out using Australian data for acidification, and global warming. Operating costs approximate 2009 US$.

### 14.5.2 *Case 2: Product Improvement through Hydrotreating of Diesel*

This case involving sulphur reduction in diesel was discussed in the context of life cycle assessment in Chapter 11. Figure 11.5 showed inventory data for a hydrotreater unit where the sulphur reduction occurs. The case has elements of micro (individual plants), meso (integrated petroleum refinery units), and macro scales (refining, distribution, and combustion of the diesel). Reduced sulphur content of diesel benefits the environment through reduced $SO_2$ emissions from transport vehicles. The extended processing system (Fig. 11.4) comprises a hydrotreater, hydrogen production, $H_2S$ recovery, and sulphur production unit supported by utility systems. Hydrogen may be available as a by-product from a catalytic reformer unit in the refinery, or from a dedicated hydrogen plant typically using steam reforming of natural gas.

**Using benefit and burden analysis**

**Environmental benefits**

- Reduced sulphur content of diesel, leading to reduced $SO_2$ emissions in combustion and hence reductions in acidification potential and toxicity
- Other benefits from combustion of cleaner diesel

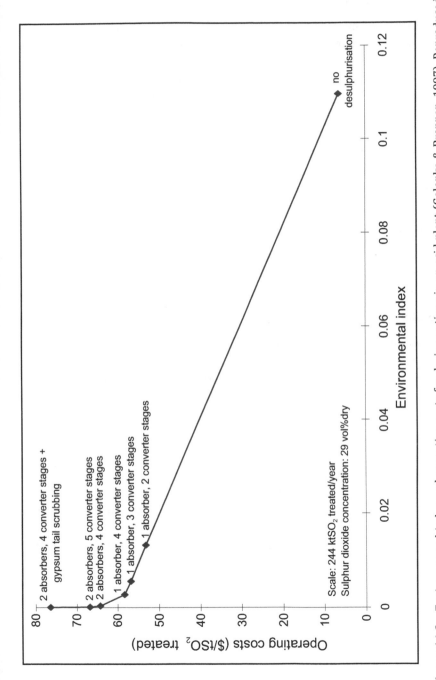

**Figure 14.2** Environmental index and operating costs for design options in an acid plant (Golonka & Brennan, 1997). Reproduced with permission from IChemE.

**Environmental burdens**

- Increased hydrocarbon emissions (predominantly fugitive) due to increased refinery processing, leading to increase in photochemical smog
- Increased energy consumption (predominantly through utility consumption) in processing, leading to increased global warming potential and increased resource depletion, assuming energy is derived from fossil fuel

**The environmental benefits $\longleftrightarrow$ burden balance** depends on the sulphur content of straight run diesel fed to hydrotreater. Feedstocks with lower sulphur contents while involving less sulphur removal to achieve the required diesel quality, still involve large energy consumption in processing.

**Economic burdens**

- Increased capital and operating costs. The hydrotreating step is a major capital cost and operating cost burden. Utilities consumptions are major contributors to operating costs

**Trade-offs**

- *Environmental*: Reduced impacts in diesel combustion versus increased impacts in diesel processing
- *Enviro versus economic*: Reduced impacts in diesel combustion versus increased capital and operating costs.

Figures 14.3 and 14.4 show plots of capital costs and operating costs incurred versus environmental impact score, calculated as the sum of normalised scores for contributing impact categories. Main contributing environmental impacts are energy depletion, and photochemical oxidant, global warming, and acidification potentials. Capital costs are based on plant capacity equivalents for processing 5 million bbl/yr of straight run gas oil. Costs approximate US$2009. These plots show the hydrotreater as the dominant contributor to cost (capital and operating) as well as to environmental impact. The sulphur recovery plant shows a negligible environmental impact because of the credit given to energy recovery on that plant.

### 14.5.3 *Case 3: Power Generation from Fossil Fuels: CCGT-NG versus ST-Br Coal*

We examine the case of an electricity generation plant using natural gas combined cycle gas turbine (CCGT-NG) compared to one using brown coal

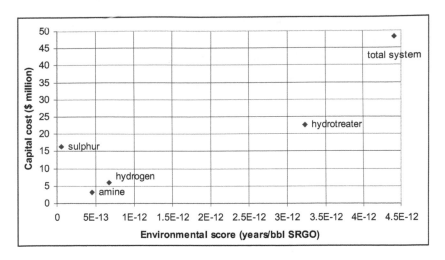

**Figure 14.3** Capital cost versus normalised environmental score — $H_2$ from dedicated plant (Burgess & Brennan, 2001). Reproduced with permission from Elsevier.

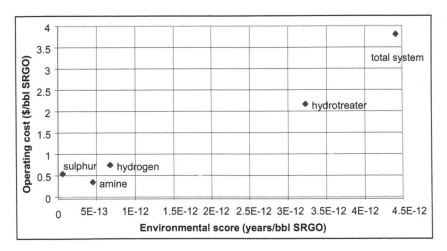

**Figure 14.4** Operating cost versus normalised environmental score — $H_2$ from dedicated plant (Burgess & Brennan, 2001). Reproduced with permission from Elsevier.

steam turbine (ST-BrC) in Australia. The environmental benefits of the natural gas power station are reduced emissions in almost all emission species and particularly carbon dioxide, see Table 11.1 in Chapter 11. The economic benefits of the natural gas power station are reduced capital and fixed operating costs (May & Brennan, 2006). The burdens of the natural gas power station are mainly increased fuel price; however, this cost burden is reduced relative to the ST-BrC case if carbon taxes are applied.

### Economic benefits of CCGT-NG versus ST-BrC power generation

- Lower capital cost
- Shorter construction time, enabling more rapid response to capacity demand changes
- Higher ratio of variable to fixed operating costs enabling viability of smaller scale power stations
- Gas power station can be located away from the fuel resource while brown coal power station must be at the source of the resource due to the high moisture content of the brown coal. This has advantages for gas power station in terms of potential access to

  ▶ electricity consumers
  ▶ $CO_2$ sequestration site (if applicable)
  ▶ water sources for cooling, boiler water feed, other demands.

### Economic burdens of CCGT-NG versus ST-BrC power generation

- Natural gas is more expensive than brown coal. Brown coal in this case has a high (>50% w/w) moisture content, effectively conferring a zero opportunity cost since coal transport costs would be prohibitively high.

### Environmental benefits of CCGT-NG versus ST-BrC power generation

- Reduced emissions per MWh for most environmentally harmful species
- Reduced fuel consumption per MWh
- Lower global warming potential per MWh.

### Environmental burdens of CCGT-NG versus ST-BrC

- More rapid resource depletion rate (proven reserves of brown coal exceed those of natural gas in Australia)
- Increased emissions of hydrocarbons.

### Trade-offs for CCGT-NG versus ST-BrC

- *Environmental*: Reduced global warming, acidification potential and other impacts versus increased resource depletion for the gas option

- *Economic*: Reduced capital costs versus increased operating costs, derived from fuel costs, for the gas option.

The relativity in operating costs is modified when a carbon tax is applied, favouring the CCGT-NG plant. A Aust\$40/t $CO_2$ tax increases the operating cost of power by 36% for the CCGT-NG plant, but by 80% for the ST-BrC plant. The carbon tax also reduces the scale penalty for smaller power stations, the penalty reduction being greater for more capital intensive brown coal plant. For a Aust\$40/t $CO_2$ tax, the operating cost scale penalty incurred by changing from a 1000 MW to 500 MW power station is reduced from 19.3% (without tax) to 10.6% (with tax) for the ST-BrC plant; the corresponding reduction for the CCGT-NG plant is from 8.2% (without tax) to 6.0% (with tax).

The comparison of the two systems can be extended to include flue gas treatment for $CO_2$ capture, pipeline transport, and sequestration. In this case, global warming potential is reduced due to $CO_2$ sequestration, but resource depletion increases due to increased energy consumed in $CO_2$ capture and compression, assuming this energy is provided by the power station generating the flue gas.

## 14.6 Environmental Effects of Scale of Production

Reductions in capital costs per unit of production capacity can be achieved through larger scale plants. Reductions in operating costs per tonne of product, as well as profitability improvements can also be achieved through larger scale plants, provided markets for products and plant capacity utilisation can be sustained. Scale economies were discussed in Chapter 13.

In contrast to economics, the effect of production scale on environmental performance has had less scrutiny. Impact category scores in LCA methodology are reported per functional unit; thus, global warming potential is reported in tonnes equivalent of carbon dioxide per tonne of product. Thus, we might conclude that environmental burdens are proportional to plant production rate and are larger for larger plant capacities. There are aspects which need closer examination, however.

The first aspect is where fugitive emissions are dominant in life cycle inventories. This has often been true of hydrocarbon emissions in petroleum refineries. For single-stream plants, although larger capacity plants may have marginally larger seals and valves, the increased leakage from such sources is likely to be substantially less than the corresponding increase in plant capacity. For parallel streamed plants, however, the number of emission points will be substantially increased for larger capacity plants leading to the potential for increased fugitive emissions. Thus, the dependence of fugitive emission inventories on plant scale may be *weak for single-stream plants* but *strong for parallel streamed plants.*

The second aspect relates to the potential for larger capacity plants, because of their lower capital and fixed operating costs per tonne of product, to invest additional capital in initiatives to reduce emissions. Such initiatives could involve feedstock purification, waste reduction at source, recycling, increased process integration, or additional effluent treatment.

The third aspect relates to the potential for increased production to result in increased emissions to the point where acceptable threshold limits in the quality of local air, water, or land are exceeded. These limits would correspond to more local impacts such as acidification or photochemical smog rather than global impacts such as global warming or ozone layer depletion. Other effects could include depletion of a locally available resource or impacts on local water receiving systems. Similar concerns occur where process plants are added to a complex (mineral, hydrocarbon, or chemical processing) on a given site.

The fourth aspect relates to the possibility that regulatory authorities may demand lower emissions per unit of product for larger capacity plants to maintain quality standards for air, water, or land in the vicinity of the plant.

Thus, we can conclude that the effects of plant scale options on environmental impacts demand rather closer scrutiny than they have received thus far.

## 14.7 Sustainability Assessment and Sustainability Metrics

Sustainability assessment covers the complete spectrum of economic, environmental, and social criteria.

Sets of indicators have been proposed to quantify sustainability impacts. Indicator sets are typically flexible, allowing indicators to be added or removed as appropriate for a particular case. Standard sets of indicators for industrial systems have been suggested (e.g., Azapagic & Perdan, 2000; Global Reporting Initiative, 2002; Institution of Chemical Engineers, 2002) although no system is universally accepted at this stage. Sustainability reporting by operating companies is discussed in Chapter 18 in the context of operations.

Considering the case of the IChemE 2002 metrics, the sustainability indicators are presented in three groups: environmental, economic, and social. Environmental indicators are listed under resource usage (covering energy, materials, water, and land), and emissions, effluents, and waste. Emissions, effluents, and waste are then further classified as atmospheric, aquatic, and land impacts. These impacts are expressed in units of tonnes of equivalent species for the specific damage category per unit of value added (e.g., global warming potential in tonnes/yr $CO_2$ equivalent per unit of added value).

Economic indicators are classified into groups as Group 1 (profit, value, and tax) and Group 2 (investments). Value added, included in the economic indicators, is the difference in value between sold products and the materials purchased to make those products. Capital inclusive value added is a similar measure to value added, but incorporates an allowance for incurred capital expenditure.

Social indicators are classified into those relating to the workplace (encompassing employment, and health and safety) and those relating to society. Society indicators include extent of meeting with external stake holders, indirect benefit to the community, number of complaints, and number of legal actions, all per unit of value added.

As demonstrated in Chapter 12, a critical component of the social criteria, particularly relevant to the process industries, is human safety. Ensuring human safety in the process industries depends on a wide range of factors encompassing the technology employed and achievement of standards over the life cycle of a given process plant in design, construction, operation, maintenance, and final decommissioning. Safety is also a key concern within the life cycle of a product, from resource extraction through to product use and final disposal.

## 14.7.1 *Case Study on Sustainability of Electricity Generation*

A study on sustainability of electricity generation from Australian fossil fuels (May & Brennan, 2006) used a range of sustainability indicators listed in Table 14.2. Employment and safety related indicators were selected as social indicators because they were relatively easy to quantify. The study included power cycles based on brown coal, black coal and natural gas. Export of Australian black coal and LNG to Japan for power generation in Japan were considered as additional cases. System boundary (Fig. 14.5) incorporated fuel extraction, fuel transport to power station, generation of power, and transmission of electricity to point of use. Note that system

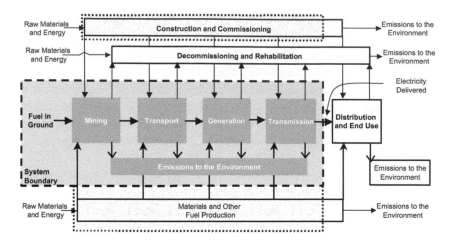

**Figure 14.5** Electricity supply system based on fossil fuels (May & Brennan, 2006). Shaded areas indicate inclusion (grey = all impacts, spotted = economic impacts only). Reproduced with permission from IChemE.

boundary considerations were judged to be slightly different for economic and environmental assessments, reflecting the relative importance of sub-system contributions for each criterion. Table 14.2 lists the values of normalised sustainability indicators obtained from the study.

In this study, no one fuel or technology system was found superior or inferior for every indicator. However, generalisations could be made: natural gas combined cycle system had advantages for most environmental and economic indicators, brown coal had an advantage in terms of value added, and black coal has a relatively poor safety performance derived from coal mining activities. Further details of the sustainability indicators and their evaluation are available in May and Brennan (2006).

## 14.8  Perception and Assessment of Risk

Risk assessment is established as a key aspect of both technical safety (see Chapter 12) and economic analysis of projects (see Chapter 13). In economic evaluation of investment projects, profitability is assessed from forecast cash flows over project life. Inbuilt into these forecast cash flows are forecasts of capital expenditures, sales volumes, selling prices, plant performance parameters, and a wide spectrum of unit costs. Risk of one or more of these forecasts falling below expectation is examined by sensitivity analysis and risk assessment. In sensitivity analysis, the effect of deviations

Table 14.2 Normalised sustainability indicators for power generation systems (May and Brennan, 2006). Reproduced with permission from IChemE)

| Indicators adopted | Units | BrC-ST | BIC-ST | IGCC-BrC | IGCC-BIC | CCGT-NG |
|---|---|---|---|---|---|---|
| **Environmental** | | | | | | |
| Resource depletion | yr/MWh × $10^{-9}$ | 0.00458 | 0.0192 | 0.00605 | 0.0134 | 0.0986 |
| Climate change | yr/MWh × $10^{-9}$ | 2.54 | 2.22 | 1.57 | 1.43 | 0.959 |
| Acidification | yr/MWh × $10^{-9}$ | 0.904 | 1.93 | 0.694 | 0.0493 | 0.120 |
| Nutrification | yr/MWh × $10^{-9}$ | 0.635 | 1.27 | 0.0709 | 0.0758 | 0.290 |
| Photochemical smog | yr/MWh × $10^{-9}$ | 0.150 | 0.318 | 0.106 | 0.0195 | 0.0994 |
| Solid waste | yr/MWh × $10^{-9}$ | 1.03 | 4.12 | 0.644 | 2.97 | 0.0007 |
| **Economic** | | | | | | |
| Value added | yr/MWh × $10^{-10}$ | 3.38 | 2.52 | 3.24 | 2.98 | 2.7 |
| Annualised cost | yr/MWh × $10^{-11}$ | 4.34 | 6.26 | 3.02 | 3.78 | 3.24 |
| Capital investment | yr/MWh × $10^{-8}$ | 9.81 | 10.0 | 5.59 | 6.03 | 4.03 |
| Capital inclusive value added | yr/MWh × $10^{-10}$ | 1.98 | 0.745 | 2.91 | 2.91 | 2.60 |
| **Social** | | | | | | |
| Direct employees | yr/MWh × $10^{-11}$ | 0.415 | 0.899 | 0.324 | 0.641 | 0.474 |
| Indirect employees | yr/MWh × $10^{-11}$ | 2.37 | 5.03 | 1.69 | 3.41 | 2.91 |
| Lost time injuries | yr/MWh × $10^{-11}$ | 0.259 | 3.35 | 0.201 | 1.96 | 0.449 |
| Fatalities | yr/MWh × $10^{-11}$ | 0.211 | 2.73 | 0.211 | 1.56 | 0.855 |

**Notes:** Values for normalised indicators have units of impact per MWh of electricity delivered per unit of impact in Australia per annum. Further information and data on impact scores prior to normalisation, contributions to impact scores from subsystems, data used for normalisation, and details of data sources can be found in May (2004) and May and Brennan (2006).

from forecast inputs is examined. The resultant NPV can be calculated and probabilities of such deviations incorporated to make an assessment of risk for the project. Such risks are generally termed business risks and can be approached systematically at the investment stage of a project.

A separate financial risk occurs where the cumulative cash flow for a project reaches such negative proportions, that a company has difficulty in meeting its interest and loan repayment commitments on borrowed capital to the point where the company's financial survival is threatened.

Unforeseen environmental damage can also result from deviations from expectations. A range of abnormal events was examined in Chapter 9 in terms of departures from design or flow sheet conditions in process plants, as well as in mining and extraction of raw materials, and transport and storage of materials. Cases of environmental damage resulting from major industrial accidents were cited in Chapter 12.

Failure to consult adequately with stakeholders can also result in less favourable project outcomes, or delays and obstacles in implementing projects. Sharratt and Chong (2002) discuss the roles of a range of stakeholder groups in capital investment decisions with emphasis on environmental risks.

Stakeholder groups identified by Sharratt and Chong as having key interests in the early stages of capital investment are

- employees and management
- shareholders
- lenders
- the national government and local authorities
- the public (including local residents, environmental pressure groups, and the general public).

The roles and perceptions of these various groups have been discussed by Sharratt and Chong in the context of a PVC manufacturing case study.

Current examples of conflict between stakeholders requiring resolution exist in the exploration and proposed extraction of coal seam methane gas in Queensland, Australia. There is conflict between advantages of tapping a valuable resource which would supplement natural gas, and the impact on traditional landowners with predominantly agricultural interests in terms of property rights infringement, potential leaks from gas wells, water contamination, and a range of environmental and safety issues.

## 14.9  Scenario Analysis

Scenario analysis can be used as an extension of sensitivity analysis in economic evaluation of projects, where the impact of various future events and time dependent trends can be explored. These might include, for example,

- difficulties in plant construction or commissioning delaying product sales realisation
- erosion in product selling price (in real terms) over time due to decline in technological competitiveness
- changes in currency exchange rates or government policy on taxation.

Scenario analysis can also be explored in terms of environmental and sustainability outcomes. Darton (2003) has explored the scenario planning technique in the context of future energy supply sources and associated emissions and atmospheric concentrations of $CO_2$. Darton argues the scenario planning technique provides an important opportunity to test current planning and policy decisions against possible future conditions.

A further role for scenario analysis is in exploring the effects of progressive tightening of environmental legislation or progressive increase in cost of emissions, for example, through a carbon tax. The effects of such changes for current operating plants and proposed projects could be explored.

## 14.10  Concluding Remarks

Sustainability assessment involves the evaluation of economic, environmental, and social credentials of existing practices and proposed improvements to processes and products. In the process industries, safety evaluation is an essential aspect of addressing social credentials. The assessments of these multiple criteria must be integrated in decision making. A systematic approach to integrating economic and environmental assessment has been discussed. This approach involves identifying and quantifying economic and environmental benefits and burdens of alternative processes, new initiatives, and incremental improvements. The approach has been illustrated for the cases of sulphuric acid manufacture, diesel desulphurisation and electric power generation. In making changes to existing technologies and practices, there is a spectrum of risks due to unforeseen outcomes which must be explored and evaluated.

Sustainability assessment is an integral part of implementing sustainability within the planning, design, and operations activities of the process industries. The application of sustainability assessment in these contexts is discussed in Chapters 15, 16 and 17.

## References

Azapagic, A. and Perdan, S. (2000) Indicators of sustainability for industry: a general framework, *Trans. IChemE.,* **78**(Part B), 243–261.

Burgess, A. A. and Brennan D. J. (2001) Desulphurisation of gas oil: case study in environmental and economic assessment, *J. Clean. Prod.*, **9**(5), 465–472.

Darton, R. C. (2003) Scenarios and metrics as guides to a sustainable future. The case of energy supply, *Trans. IChemE.*, **81**(Part B), 295–302.

Global Reporting Initiative. (2002) Sustainability Reporting Guidelines, Boston, United States. http://www.globalreporting.org/NR/rdonlyres/529105CC-89D8-405F-87CF-12A601AB3831/0/2002_Guidelines_ENG.pdf

Golonka, K. A. and Brennan, D. J. (1997) Costs and environmental impacts in pollutant treatment. A case study of sulphur dioxide emissions from metallurgical smelters, *Trans. IChemE.*, **75**(Part B), 232–244.

Institution of Chemical Engineers. (2002) The Sustainability Metrics, IChemE http://www.icheme.org/sustainability.

May, J. (2004) Sustainability of Electricity Generation using Australian Fossil Fuels, PhD thesis, Monash University, Melbourne.

May, J. R. and Brennan, D. J. (2006) Sustainability assessment of Australian electricity generation, *Trans. IChemE.*, **81**(Part B), 131–142.

Sharratt, P. N. and Chong, P. M. (2002) A life cycle framework to analyses business risk in process industry projects, *J. Clean. Prod.*, **10**, 479–493.

# Problems: Part C

## 1. Life Cycle Assessment — Electricity Generation in Australia

Much of the electric power generated for industrial use is obtained from black coal using steam and power cycles. The composition of a black coal supplied to a power station is given as follows:

   10.0 wt% water
   16.7 wt % ash
    0.5 wt % sulphur
   60.0 wt % carbon
    8.0 wt% oxygen
    3.6 wt% hydrogen
    1.2 wt % nitrogen

The efficiency of power generation, defined as the ratio of power generated to fuel energy consumed, may be taken as 36%, corresponding to a heating value of 24 GJ per tonne coal.

(a) Calculate the consumption of coal and the emissions of carbon dioxide and sulphur dioxide in the power station per MWh of electricity generated. Assume no flue gas desulphurisation is performed in the power station.

(b) Place the resources consumed and gaseous wastes generated in the combustion of black coal in the appropriate impact categories under life cycle assessment classification.

(c) If the emissions of nitrogen oxides from fuel and thermal sources amount to 2.2 kg of $NO_x$ per MWh of electricity generated, calculate the normalised impact score for acidification potential derived from oxides of sulphur and nitrogen, including its units.

(d) Identify the main contributions to water consumption and waste water generation in the steam and power cycles.

(e) In taking a life cycle view of 1 MWh of electricity generated at the power station and delivered to an industrial consumer, identify stages of the life cycle which should be considered beyond generation.
(f) Explain what is meant by the term 'external costs' or externalities. Give examples of externalities derived from environmental impacts in power generation from black coal.
(g) Identify the major sustainability concerns regarding electricity generation from black coal, giving your reasons.

## 2. Life Cycle Assessment: Natural Gas Purification Plant

A purification plant is proposed for a natural gas field located in Victoria, Australia. The gas field has a negligible amount of hydrocarbons present in the liquid phase. A Life Cycle Assessment study has resulted in an estimate of the inventory data for the proposed plant. The inventory data account for the process, subdivided into flares and vents, as well as for utilities used on the plant (mainly electricity and hot and cold utilities). Inventory data are expressed in kg per GJ of sales gas produced. Some sour water streams are also anticipated, but can be neglected for the purpose of this exercise.

**Inventory Data**
**Input**

| | |
|---|---|
| Coal: | 0.88 |
| Water: | 7.4 |
| Natural gas: | 1.73 |

| Output | Process Flares | Process Vents | Utilities |
|---|---|---|---|
| Methane | | 0.051 | |
| Non methane hydrocarbons | | 0.018 | |
| Carbon dioxide | 16.7 | 1.04 | 7.31 |
| Sulphur dioxide | 0.0046 | | 0.0092 |
| Nitrogen oxides as $NO_x$ | 0.0060 | | 0.010 |
| Hydrogen sulphide | | 0.0004 | |
| Ash | | | 0.14 |

**Further Data**
**A. Estimates of world reserves (tonnes)**

| | |
|---|---|
| Coal: | $2.10 \times 10^{12}$ |
| Water: | $4.00 \times 10^{13}$ |
| Natural gas: | $7.80 \times 10^{10}$ |

**B. Equivalency factors for Life Cycle Assessment classification**

Methane compared with carbon dioxide for global warming potential may be taken as 21.

$NO_x$ compared with $SO_2$ for acidification may be taken as 0.7

$H_2S$ compared with $SO_2$ for acidification may be taken as 1.88

**C. Estimates of Australian consumption or generation**

| | |
|---|---|
| Coal consumption | $1.91 \times 10^8$ tonnes/yr |
| Natural gas consumption | $1.49 \times 10^7$ tonnes /yr |
| Water consumption | $1.46 \times 10^{10}$ tonnes/yr |
| Solid waste generation | $6.60 \times 10^8$ tonnes/yr |
| Global warming | $4.03 \times 10^8$ tonnes of $CO_2$ equivalent per year |
| Acidification | $2.94 \times 10^6$ tonnes of $SO_2$ equivalent per year |

(a) Use the data provided and the classification and normalisation pro-cedures of life cycle assessment methodology to calculate *normalised impact scores* for the following environmental impact categories. Give the units for each normalised score:

   (i) resource depletion

   (ii) global warming potential

   (iii) acidification potential.

(b) What conclusions can be drawn from the normalised impact scores calculated in part (a)?

(c) List two additional impact categories to which some inventory data could be classified. For each impact category, nominate the molecular species contributing to that impact.

(d) Outline some key considerations necessary in arriving at a reliable estimate of the inventory data at the design stage of a gas purification plant of this type.

(e) List two additional wastes you would expect from the operation of a natural gas purification plant which are not accounted for in the inventory data provided. Identify the sources of these wastes.

(f) The natural gas from the field is intended for supply to a power generation plant located in the adjoining state of South Australia. In accounting for environmental impacts of gas supplied to the power station gate, what additional systems other than the purification plant should be considered? Provide a simple diagram and give reasons to support your answer.

(g) In the event of such a gas field having considerably larger quantities of hydrocarbon liquids in the gas, nominate two bases which could be considered for apportioning environmental burdens between the product gas and the product liquid streams. Briefly explain the merits or otherwise of each basis.

## 3. Sea Water Desalination

Due to shortages of good quality fresh water in many cities, towns, and remote industrial sites, interest has grown in the potential for sea water desalination as a source of fresh water. Initiatives to build new desalination plants have caused controversy, not only because of impacts at proposed sites, but also in relation to costs and energy requirements. In this problem, we consider a proposed desalination plant near Wonthaggi, Victoria for supplying fresh water to Melbourne. The plant uses reverse osmosis technology. Some preliminary cost and performance estimates are given in an executive summary of a seawater desalination feasibility study by GHD P/L and Melbourne Water.

### (a) Capital Cost Estimation

An estimate of the fixed capital cost of the project for a proposed plant of 150 Gl/yr capacity was estimated at $3.1 billion (in 2007 $ Australian). Of this total cost, the cost of the treatment plant for converting sea water to fresh water was $1.16 billion (2007 $ Australian). Other components of the project cost included intake and outfall tunnels for the plant, transfer pipelines for desalinated water, and construction and management costs. For the treatment plant,

(i) nominate the main equipment item categories you would expect
(ii) estimate the total purchased cost of the equipment items.

### (b) Energy requirements

(i) Energy is required for both the treatment and transport of water. The power requirements for the 150 Gl capacity plant was estimated at 90 MW. Make an approximate estimate in $kWh/m^3$ of the total power requirements for the proposed Victorian plant. Compare this estimate with published performance data for the Larnaca plant in Cyprus.

### (c) Operating cost estimation

(i) Prepare an operating cost sheet for desalinated water based on energy and capital dependent costs. Include in the cost sheet columns for

- cost category
- consumption of energy
- unit costs of energy (assume $80/MWh)
- basis for estimation of fixed operating costs
- annual cost
- cost per m$^3$ desalinated water.

(ii) What additional operating costs might be incurred beyond the energy and capital related costs estimated in (i)?

**(d) Economic viability**

Based on your capital and operating cost estimates, estimate a minimum viable selling price for water delivered to consumers. Present your calculations and assumptions concisely.

**(e) Wastes**

Make a list of the major wastes, other than those related to energy consumption, which would result from the plant's operation.

**(f) Technology**

Give a concise explanation of the principles of the reverse osmosis technology. What alternative technology to reverse osmosis is available for desalination plants? Under what circumstances might this alternative technology be preferred to reverse osmosis?

## References

GHD and Melbourne Water (June 2007) Seawater Desalination Feasibility Study, Executive Summary. http://www.melbournewater.com.au/content/library/current_projects/water_supply/seawater_desalination_plant/feasibility_study/Desalination_plant_Feasibility_Study_Executive_Summary.pdf

Koutsakos, E. (2008) Thirsting for chemical engineers, *The Chemical Engineer*, June, pp. 26–27 in relation to Lanarca desalination plant, Cyprus.

Larnaca Sea Water Reverse Osmosis (SWRO) Plant, Cyprus http://www.water-technology.net/projects/larnaca/

## 4. PVC Production: Water Consumption and Carbon Dioxide Emissions over Cradle to Gate Life Cycle

Consider the cradle to gate portion of the product life cycle for PVC where the 'gate product' is rigid PVC despatched from a PVC factory *prior to conversion to plastic products*. For the main process steps in the cradle to gate life cycle, obtain literature data for raw materials and utilities consumptions per tonne of product for both PVC and upstream intermediate products. Hence make a quantitative estimate of

(a) the consumption of water per tonne of PVC product
(b) the emission of carbon dioxide per tonne of PVC product for the cradle to gate life cycle.
   Where possible, account separately for the water consumptions and $CO_2$ emissions derived from

   (i) process sources
   (ii) utility sources.

## Useful Information Sources

Kirk and Othmer (1994) *Encyclopaedia of Chemical Technology*, 4th edn, Wiley, New York.

Section 11.2 in Chapter 11 titled 'Life cycle inventories for common utilities'.

Ullmann (2002) *Encyclopaedia of Industrial Chemistry*, Wiley-VCH, Weinheim, Germany.

## 5. Sustainability of Electricity Generation

(a) From the inventory data provided in Table 1 below, compare the fuel resources consumed and the wastes generated for the following technology and fuel systems used for electric power generation:

   (i) Steam turbine — Brown coal: **ST-BrC**
   (ii) Combined cycle gas turbine — natural gas: **CCGT-NG**

   Calculate fuel consumption and $CO_2$ emissions per MWh generated for each system, and compare your calculated values with the data in Table 1.

(b) Use the fuel consumption, gaseous emission and solid waste data in Table 1 to estimate

   (i) environmental impact scores
   (ii) normalised environmental impact scores (using Australian data) for the various categories of environmental damage resulting.

(c) Use the estimates of capital and fuel costs given below to estimate the operating cost in $/MWh of delivered electricity for power stations of 500 MW and 1000 MW capacity for the BrC-ST, CCGT-NG systems

   (i) with zero carbon tax
   (ii) with a carbon tax of $40/tonne $CO_2$ emitted.

Table 1. Inventory of inputs and outputs for extended electricity supply systems (adapted from May & Brennan, 2002)

| System | ST-BrC | ST-BlC | OCGT-NG | IGCC-BrC | IGCC-BlC | CCGT-NG | CCGT-LNG |
|---|---|---|---|---|---|---|---|
| *Inputs* | | | | | | | |
| Main fuel | 1464 | 467 | 303 | 754 | 261 | 159 | 178 |
| Limestone | | | | | 35.7 | | |
| *Outputs* | | | | | | | |
| Solids | | | | | | | |
| Ash | 21.8 | 87.3 | | 6.8 | 26.7 | | |
| Gases | | | | | | | |
| $CO_2$ | 1209 | 1023 | 825 | 740 | 660 | 445 | 535 |
| $CH_4$ | 0.007 | 1.65 | 0.979 | 0.340 | 0.974 | 0.478 | 1.28 |
| NMVOC | 0.022 | 0.024 | 0.048 | 0.011 | 0.012 | 0.174 | 0.475 |
| $N_2O$ | 0.018 | 0.009 | 0.006 | 0.011 | 0.012 | 0.010 | 0.014 |
| $NO_x$ | 1.68 | 3.36 | 2.83 | 0.187 | 0.200 | 0.767 | 1.29 |
| CO | 0.223 | 0.142 | 0.685 | 0.211 | 0.167 | 0.096 | 0.228 |
| $SO_2$ | 1.91 | 4.17 | 0.045 | 1.93 | 0.059 | 0.026 | 0.100 |
| Particulates | 0.493 | 0.548 | 0.285 | 0.022 | 0.038 | 0.037 | 0.063 |

**Notes**
1. Units are in kg/MWh delivered
2. Flue gas scrubbing has been assumed for IGCC-BlC
3. The system nomenclature adopted is as follows
   ST-BrC Steam turbine using brown coal
   ST-BlC Steam turbine using black coal
   OCGT-NG Open cycle gas turbine using natural gas
   IGCC-BrC Integrated gasification combined cycle using brown coal
   IGCC-BlC Integrated gasification combined cycle using black coal
   CCGT-NG Combined cycle gas turbine using natural gas CCGT-LNG Combined cycle gas turbine using liquefied natural gas

What are the main effects of the carbon tax on the costs of power generated?

## Assume for all cases

- 30-year operating life
- fuel costs account for all variable costs
- annual fixed operating cash costs (excluding loan repayment) of 5% fixed capital
- real cost of capital (or discount rate) of 5% per annum

Capital costs (2008 $ Australian) for 1000 MW power station:

Brown coal — steam turbine   $3.8 billion
CCGT natural gas             $1.31 billion

Fuel costs delivered to power station (2008 $ Australian):
   $7/tonne brown coal
   $4/GJ natural gas.

## Reference

May, J. R., and Brennan, D. J. (2004) Life cycle assessment of Australian fossil fuel
   energy options, *Trans IChemE*, **81**(Part B), 317–330, 2003, and p. 81, 2004.

# IMPLEMENTATION

# Chapter 15

# Planning for Sustainable Process Industries

## 15.1 Introduction

Systematic planning is essential for achieving improved sustainability. Planning is necessary at many levels in industry, spanning research and development for new technologies, further development of established process technologies, design and construction, plant operations, and a range of related business activities. Planning for sustainability encompasses many divergent activities. Aspects such as adequate funds, time, personnel and technology, future demand patterns, trends in population and economic development, government policy on economic incentives and taxation, as well as environmental regulation must all be considered. Planning is also necessary at government level in developing improved regulatory frameworks and in providing incentives and support for sustainability initiatives in industrial, commercial, and domestic activities.

## 15.2 Forecasting

Forecasting is an integral part of industry planning. Market forecasting for product demand and price is essential to guide revenue estimation and plant sizing decisions for projects. Market volume for many commodity products is, in turn, influenced by population size and the economic well-being of potential customers; hence forecasting population growth and economic growth

*Sustainable Process Engineering: Concepts, Strategies, Evaluation, and Implementation*
David Brennan
Copyright © 2013 Pan Stanford Publishing Pte. Ltd.
ISBN 978-981-4316-78-1 (Hardcover), 978-981-4364-22-5 (eBook)
www.panstanford.com

becomes important. Technical performance, environmental performance, and selling price are key aspects in placing a product in its competitive niche. Technology forecasting is essential not only to guide the selection of process technologies in new projects but also to assess the competitiveness of technologies already employed by an industry.

Sigmoid (or S-shaped) curves of performance against time have been widely used for modelling both market growth and technology performance. Sigmoid curves are also a useful qualitative basis for representing the life cycle of a process technology, incorporating initiation, growth, and maturity phases.

Technology performance indicators include dominant parameters specific to a technology such as technical efficiencies, but also maximum available capacity of equipment or plant, as well as productivity performance in the key areas of raw materials and energy consumption, personnel employed, and capital investment per tonne of product.

## 15.3 Scenario Development

Projects involving process plant investment are developed based on best estimates of markets, technology performance, capital and operating costs, and project life. However, since the project life extends some 20 or 30 years beyond the time of evaluation, it is important to explore deviations from these estimates and their impacts on project viability. There are many possible deviations from estimates arising from acts of competitors, initiatives from other industries, technological change, discoveries of new raw material sources, reduced availability of raw materials, decline in raw material quality, tighter regulatory control on emissions, changes in government policy on tax, higher prices for utilities, and so on. To explore these possibilities it is necessary to postulate scenarios and explore the performance of the proposed project in the light of these scenarios.

Such scenario development is essential in the process industries not only for new projects, but also in all aspects of business planning including technology development, raw material sourcing, utilities supply contracts, waste disposal, and employee resources. Scenario development is also a key part of exploring environmental, safety, and sustainability outcomes, as discussed in Chapters 12 and 14.

Scenario development is likewise important for governments in ensuring infrastructure, water and energy supply, landfill provisions, transport facilities, and availability of skilled labour and professionals. Governments must also consider scenarios of population growth and industry development in relation to future emissions and ensuring the quality of air, water, and land and the wellbeing of communities.

## 15.4 Technology Innovation

Technology improvement in processes, products, equipment, and plant is essential to ensure business competitiveness. Improvement is necessary to meet the spectrum of criteria we have considered in Chapters 10 to 14. A great deal of focus in the process industries has recently been given to catalysis, nanotechnology, biotechnology, improved materials, and novel reactors such as microwave and ultrasonic, but there is continued need for improvement in all aspects of process technology.

In the context of population growth, improvement in resource efficiency and emissions over the life cycle of products is essential to sustain or improve environmental performance. Some specific aspects of technology innovation relevant to the process industries are now explored.

### 15.4.1 *Intensification*

Intensification has been actively pursued, especially over the last 30 years, as a means of making major improvements in technology. For equipment, intensification means an increase in capacity for a given equipment size, or conversely a decrease in physical size for a given duty. The smaller physical size offers benefits in space, weight, and capital cost (with savings in materials requirements, foundations, installation labour) and reduced inventories of process materials. Reduced inventories offer benefits in inherent safety where hazardous materials are involved. Intensification of equipment is especially important in the context of off-shore oil and gas processing.

Process intensification often stems from reactor improvements in yield and selectivity; product outputs are increased, raw materials consumptions per tonne of product are reduced, and capacity requirements for downstream separation and materials recycling are reduced. Other process intensification examples include reducing batch times, converting batch processes to continuous processes, and reducing the number of overall processing steps. An example of process step reduction is catalytic distillation where a heterogeneous catalytic reaction is combined with distillation in a single column.

Intensification is achieved through changes in technology. Much of recent research has focused on micro mixing studies in reactors, and use of centrifugal force in reaction, mass transfer, and heat transfer equipment. There has been continuing interest and activity in design of more compact heat exchangers (Wadekar, 2000) as well as gas–liquid contacting columns, for example, through the use of structured packing.

## 15.4.2 *Technology Diffusion*

Opportunities exist for advances in one scientific discipline to be applied in other scientific disciplines. Advances in materials technology, for example, have lead to major improvements in a number of process technologies during their evolution. Some instances of this are as follows:

| Process technology | Materials developments |
|---|---|
| Caustic/chlorine | titanium, polymers, reinforced polymers, membranes |
| Ethylene and vinyl chloride monomer | improved alloys permitting higher temperatures in reactors |
| Hydrotreating and catalytic cracking units in refineries | progressive improvements in catalysts |

## 15.4.3 *Technology Evolution*

The evolution of technology incurs research and development effort, time, and investment to reach a point where commercialisation can occur. Even then, further scale development and performance improvement at process plant level is usually required to make the technology competitive. Performance development is commonly modelled as an S-shaped curve comprising three phases:

- initiation, involving long periods to achieve commercialisation, technical uncertainty and uncertainty regarding commercial potential
- rapid growth, typified by scale development to achieve scale related economies in capital and operating costs
- maturity, where rate of cost reduction and scale development declines markedly; competition may also emerge from newer, superior technologies.

## 15.4.4  *Rates of Change*

Rates of change in performance for process technologies have generally been reported as equivalent exponential rates. These are approximations but serve as useful indicators of progress. Performance parameters reported include productivities in raw materials consumption, energy consumption, personnel employed, fixed capital, and production cost, as well as maximum available plant capacity. Some early performance data for two technologies reviewed in Chapter 4 are listed in Table 15.1 (Brennan, 1987, 1988). Different rates of progress often reflect introduction of new technology within the main process technology. Decline in capital productivity in the 1970s resulted from plateauing of plant scale development, as well as incremental capital expenditures to reduce feedstock and energy consumption, maximise by-product recovery, and improve reliability, safety, and environmental standards.

Table 15.1   Performance improvement over time in new plants employing best available technology, expressed as equivalent exponential rates of change (%/yr)

| Technology | Chlorine/caustic soda Mercury cell | Ethylene from steam cracking of naphtha |
|---|---|---|
| **Performance parameter** | **Improvement rate %/yr** | **Improvement rate %/yr** |
| Maximum available plant capacity t product per day (tpd) | 1940–1975: 8.7 1970–1985: 0.0 | 1950–1970: 13.0 1970–1985: 1.2 |
| Raw materials consumption t per t product | 1940–1980: 0.0 | 1950–1970: 1.9 1970–1985: 1.2 |
| Energy consumption MWh or MJ per t product | 1900–1975: 0.0 1975–1980: 3.6 | 1950–1960: 1.0 1960–1985: 4.6 |
| Personnel employed per tpd product | 1940–1975: 8.0 1975–1980: 2.0 | 1950–1969: 9.0 1969–1985: 1.4 |
| Fixed capital productivity (t product/yr$) | 1940–1972: 5.6 1972–1980: −7.2 | 1950–1966: 7.5 1966–1985: −7.4 |
| Total production cost productivity (t product/$) | 1940–1950: 3.3 1950–1960: 3.0 1960–1970: 3.2 1970–1980: 0.0 | 1950–1960: 2.5 1960–1970: 2.2 1970–1980: 0.9 |

## 15.5  Transition to Renewable Feedstocks

Increasing consumption of fossil fuels and traditional raw materials for process industries and their products has given rise to concerns over resource depletion. Processing of mineral ores to metals provides opportunities for recycling metals after their use in various product or construction applications. However, once fossil fuels are combusted, there is very limited opportunity to recycle flue gas components. In the light of these concerns there has been increasing interest in the further development of processes, products, and fuels based on renewable resources.

The term biomass has been defined by Moens (2006) as comprising all plant materials such as wood and its wastes, herbaceous and aquatic plants, agricultural crops and their processing residues, and most of the organic portion of municipal wastes. The main chemical components of biomass are cellulose, hemicellulose, and lignin, which are collectively referred to as 'lignocellulose'. The composition of biomass is complex and may vary with agricultural practices, for example, in segregating different plant components in harvesting.

Biomass has an important role in the natural carbon cycle. Solar energy is used to chemically combine atmospheric carbon dioxide with water in soil to generate plant polymers which have potential for conversion to fuels and chemical feedstocks. Thus, biomass production acts as a sink for atmospheric carbon dioxide while providing a potentially valuable resource. Biomass can be converted to distinct plant products by separation processes, to gas by microbial digestion or gasification, to liquid fuels by pyrolysis, to sugars as intermediate feedstock for ethanol, and to glycerides by extraction. Biomass has also been identified as a potential source of natural fibres for construction materials.

Increasing interest has focussed on the concept of biorefineries for the integrated processing of biomass to produce chemicals, fibres, food products, fuels, and energy. A comprehensive review of renewable feedstocks and the nature and structure of potential biorefineries is provided by Moens (2006). A number of demonstration and commercial biorefineries have been constructed in the United States (http://www1.eere.energy.gov/biomass/integrated_biorefineries.html).

There are challenges in technology in developing conversion processes and increasing their efficiency, and in ability to handle varying feedstock compositions as well as optimising the economic value of outputs. There are challenges in matching the seasonal harvesting of biomass to market demands for processed products. There are also sustainability challenges in land management to ensure that priorities in relation to food supply, energy sources, and process feedstocks are resolved.

A transition from reliance on abiotic resources to the efficient use of renewable feedstocks will take time, investment, scientific and engineering skills, and perseverance.

## 15.6  Site Selection for Process Plants

Dominant commercial driving forces for site location of process plants are access to both markets for products and feedstocks to minimise costs of transport and distribution. However, in most cases, only one of these criteria can be satisfied. There are industries where choice of plant location is resource dominated, and industries where it is market dominated.

Examples of resource located industries include the petrochemical industries based on the oil and gas supplies and the salt deposits of the Middle East. Methane (from natural gas) can be burned to generate steam and electricity, and can also be used as a raw material for making hydrogen, ammonia, and methanol; ethane (from natural gas) can be used as a feedstock for ethylene. Both methane and ethane can be priced cheaply, since they are by-products of a predominantly oil production industry; both gases have often been flared in the past. Ammonia, methanol, and ethylene are all raw materials for making a wide range of downstream chemical products. These downstream products can be made on a sufficiently large scale to enable a profitable export industry.

Likewise, the fertiliser industry can be usefully located near deposits of phosphate rock, for example, in Florida, USA, and Jordan in the Middle East. After mining and beneficiation, the phosphate rock can be processed on a large scale to phosphoric acid and downstream derivatives such as ammonium phosphates and triple superphosphate. These products are made on a large scale supported by large rock reserves, and then exported. The economic and environmental burdens are more favourable for higher value-added products than for basic raw materials; transport costs become a smaller proportion of product values and transport of low-value impurities is avoided.

Some national process industries, for example, those of Japan, have very little indigenous raw material but have a large, sophisticated industrial market locally, supported by a large and confined population. Through innovation and locally developed technology, supported by good industrial planning and management, the disadvantage of higher raw material costs are offset to enable a highly competitive industry capable of servicing both local and export markets. While raw materials must be imported incurring transport costs, Japan as a dominant world consumer of raw materials can source supplies in large quantities as a competitive, well-organised purchaser.

Large, well-integrated sites can offer a newly added plant, as well as the established plants, cheap feedstock, cheap utilities, and other benefits derived from large scale and an established industrial infrastructure. Access to shipping through port facilities is a key location advantage, and is usually part of the infrastructure of an established site. One example is in Belgium, where approximately 20,000 vessels from 70 nations enter the Antwerp port each year carrying some 100 million tonnes of freight, mainly crude oil, foodstuffs, raw materials for the chemical industry, mineral ores, and manufactured goods such as vehicles. Storage tanks with a combined capacity in excess of 10 million m$^3$ hold oil and oil-based products.

Site location affects plant capital costs. Remote sites incur capital cost penalties in plant construction because of increased freight costs for equipment and materials to the site, and the costs of attracting and accommodating the necessary workforce. Cost penalties may be incurred at some sites due to steep terrain or severe weather conditions. Many plants located at raw material sources suffer such isolation and geographical disadvantages.

Some sites are more prone to environmental problems than others. This may be related, for example, to the sensitivity of the landscape (e.g., forest or desert) to certain emissions, or lack of space for solid waste disposal. There are also more stringent government environmental regulations in some countries than others, as well as the effects of different background pollution levels in certain regions, including highly populated cities.

The full spectrum of operating costs including materials transport can depend on location. Location influences the costs of employing personnel

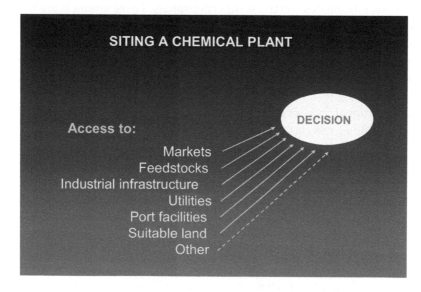

**Figure 15.1** Important considerations in siting a chemical plant.

for operations. Raw materials and utilities costs can be sensitive to location reflecting the availability and quality of key resources, as well as scale of supply or generation.

Scale of process plants can be influenced by location. Larger scale plants are usually feasible where there is abundant feedstock supply, and on a large integrated complex, where an upstream plant supplies intermediate chemicals to several downstream plants.

Superimposed on any natural advantages of industries and their national economies are the various government policies relating to business and industry. Many governments are prepared to offer tax concessions to encourage industry development.

Ultimately, a location decision for a plant is decided by weighing up many factors, with some key aspects depicted in Fig. 15.1.

## 15.7 Integration of Process Plants and Process Industries

The structure of a process industry can be defined, in part, according to the linkages between component plants, and the material and energy flows between the plants. Process plants can be integrated on the one site offering a range of economic, safety, and environmental advantages. Plants on different sites can also be linked to improve the industry ecology as discussed in Chapter 3. The principles of industrial ecology should be integrated with site location decisions as part of industrial planning.

Van Beers, Corder, Bossilkov, and van Berkel (2007) have identified drivers, barriers, and 'triggers' for regional synergies based on studies of industrial development at Kwinana (Western Australia) and Gladsone (Queensland) in Australia. The drivers, barriers and triggers have been explored under the categories of

- economics
- information availability
- corporate citizenship and business strategy
- region specific issues
- regulation
- technical issues.

Some potential barriers derive from safety implications and operating complications of transferring process and energy streams between different operating sites, and the possibility of intermittent stream flows with variations in flow rate and composition.

Industry structure can also be characterised according to the ownership, capital financing, and management of the component plants. In this context, industries can be integrated both *horizontally* and *vertically*. Companies can integrate horizontally by increasing the number of similar operations — for example, the union of multiple ethylene manufacturing plants or the union of multiple dairy processing centres. The main purpose of horizontal integration is to achieve business economies of scale.

Companies can be integrated *vertically* to become involved in upstream or downstream business activities. Vertical integration in manufacturing can be

- downstream (or forward) towards increased control of its markets
- upstream (backward) towards increased control of its raw materials supplies.

Thus, an ethylene producer can integrate downstream by acquiring a vinyl chloride monomer plant. A PVC manufacturing plant could integrate both downstream into PVC pipe fabrication and upstream into vinyl chloride monomer manufacture. Regular review of the extent of integration and its merits is an essential part of business planning.

## 15.8 Distributed Manufacture

Benson and Ponton (1993) in their pioneering paper on process minia-turisation argued for smaller scale manufacture of hazardous chemicals at their point of use, rather than large-scale centralized production with subsequent transport. Such transport often involves large inventories of hazardous materials being moved through urban roadways and population centres. Transport requires materials storage at product despatch and arrival points to correspond with the batch nature of the transport. There are risks of loss of containment in such transport and storage, exacerbated by the magnitude of inventories involved.

Key characteristics envisaged by Benson and Ponton for miniaturised distributed plants include

- minimum inventories of feedstocks, products, and materials in process
- small and intensified processes and equipment
- flexibility to operate over a wide capacity range
- a high degree of automation.

Inherent in the arguments of Benson and Ponton was the need for scaled-down reactors, possibly including novel design aspects, as part of the miniaturized processes. Case examples considered were chlorine, methanol, and hydrogen cyanide (HCN) production.

The concept of distributed manufacture of chemicals, electricity, and fuels (especially hydrogen) as an alternative to centralized production and distribution is increasingly suggested today in discussions on possible future technologies.

### 15.8.1 *Case of Aqueous Sodium Cyanide Production at the Point of Use*

This case study is provided as an example of the potential for distributed manufacture. Some driving forces, implications for technology selection and process integration are discussed.

In 2002, undergraduate chemical engineering students at Monash University addressed small-scale production of aqueous sodium cyanide at Australian gold processing centres for their design project. In all cases, access to piped natural gas supply was assumed. Design teams were assigned specific plant capacities and locations, mainly in Western Australia but were free to choose the process route. This case and some of its implications has been discussed by Brennan and Hoadley (2003).

Prior to 1990, sodium cyanide was imported into Australia as a solid. Expansion of Australian gold production, triggered by an increase in the world price for gold, resulted in an increase in imports of cyanide into Australia from 2000 tonnes in 1975 to 52,000 tonnes in 1988. This growth in imports led to the construction of three sodium cyanide plants in 1990:

- two in Gladstone, Queensland, using different processes to produce solid
- one in Kwinana, Western Australia, producing aqueous product for rail transport in special containers.

Since 1990, further expansion of the Australian gold industry has led to expanded production of sodium cyanide within Australia. The Kwinana plant has also been modified to produce solid as well as aqueous product.

Demand for sodium cyanide in Australia is widely dispersed, but predominates in Western Australia. This contrasts with a concentration of production capacity in Gladstone. Largest demands are near Kalgoorlie, but large potential markets also exist at Telfer and Boddington.

Sodium cyanide manufacture in Australia involves high temperature synthesis of HCN, rapid cooling of reactor gases in a waste heat boiler, and absorption of HCN in caustic soda. There are three main options for HCN synthesis (Hoffmann, Hungerbuhler, and McRae, 2001; Kirk and Othmer, 1994; Ullmann, 2002):

- Reaction for the *Andrussow Process* (discussed in Chapter 4) involves **exo**thermic oxidation of methane and ammonia at 1100°C over a platinum–rhodium catalyst.

$$CH_4 + 1.5O_2 + NH_3 \quad \rightarrow HCN + 3H_2O - 481.9\,kJ/mol$$

- Reaction for the *Degussa-BMA* process occurs at a slightly higher temperature, but is **endo**thermic and occurs in the absence of air. Ceramic reactor tubes coated with a platinum based catalyst are housed within a furnace.

$$CH_4 + NH_3 \quad \rightarrow HCN + 3H_2 + 252\,kJ/mol$$

- Reaction for the *Shawinigan process* occurs between ammonia and propane (or butane or natural gas) in the absence of air in a fluidized bed of coke heated electrically to 1500°C. This reaction is **endo**thermic.

$$3NH_3 + C_3H_8 \quad \rightarrow 3HCN + 7H_2 + 634\,kJ/mol \qquad (15.1)$$

The ensuing absorption of HCN in caustic soda is exothermic, and may be accompanied by sodium carbonate formation if carbon dioxide is present in the reactor effluent gas.

Because of the toxicity of cyanide, Government authorities have permitted transport of aqueous sodium cyanide only for specific routes, mainly by rail over restricted distances and in specially designed containers. The solid product has been vastly preferred because of greater ease of containment in the event of spillage during transport. Production of solid sodium cyanide compared with aqueous product (30% w/w NaCN) incurs additional capital and operating costs associated with evaporation, crystallization, drying, briquetting, and packaging (Fig. 15.2). The packaging operation is labour intensive. NaCN purity demands are more stringent for solid than aqueous product to ensure satisfactory performance of crystallization, drying and briquetting steps.

Table 15.2 shows indicative post-reactor concentrations for the three main HCN process options, based on estimates drawn from Ullmann (2002) and Kirk and Othmer (1994). While the Andrussow process has been preferred by Australian producers, comparison of reactor outlet

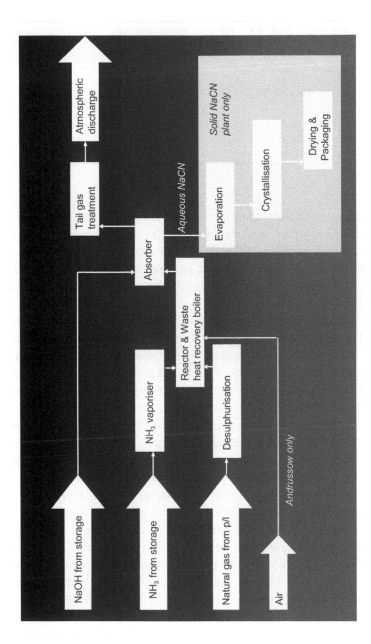

**Figure 15.2** Process routes for aqueous and solid NaCN products. (**Note:** p/l denotes pipeline.)

Table 15.2    Indicative post reactor gas concentrations for alternative HCN production process routes

| Compound | Andrussow (mol %) | Degussa-BMA (mol %) | Shawinigan (mol %) |
|---|---|---|---|
| HCN | 7.8 | 22.9 | 25.0 |
| $NH_3$ | 2.3 | 2.5 | 0.005 |
| $H_2$ | 15.0 | 71.8 | 72.0 |
| $N_2$ | 49.0 | 1.1 | 3.0 |
| $CH_4$ | 0.5 | 1.4 | |
| CO | 3.9 | | |
| $H_2O$ | 21.1 | | |
| $CO_2$ | 0.4 | | |

gas compositions indicates some potential benefits of the BMA process (Fig. 15.3). Gas volumes are smaller for the BMA process than the Andrussow process because of the absence of nitrogen and water vapour, HCN concentrations are higher, and tail gas leaving the absorber is richer in hydrogen. The higher hydrogen concentration could be an advantage if hydrogen can be profitably utilized.

Reactor concepts for the Andrussow and BMA routes are different (Fig. 15.3). The Andrussow route resembles ammonia oxidation in nitric acid plants, with the platinum catalyst gauze sitting directly above a waste heat boiler. The traditional BMA reactor resembles a hydrogen reformer furnace. More recent Degussa patents describe a more compact and energy efficient design which could benefit small-scale applications.

While the Andrussow process appears favoured by access to know-how and by operability and maintenance characteristics, the BMA process appears to offer advantages in raw material efficiency, in potential for some capital savings, and in providing by-product hydrogen.

One issue for remote plants is the use of steam generated in the waste heat boiler following reaction. Steam is beneficial only if it can be used effectively. There is little demand for steam within aqueous sodium cyanide production. The BMA process generates less steam than the Andrussow process per tonne of product. An alternative to the waste heat boiler could be direct water quenching; in this case, the resultant dilution of the sodium cyanide solution would be less for the BMA process, and acceptable for use by the gold producer.

Opportunities for process integration exist at both *micro* and *macro* levels. At micro level there are opportunities in $CO_2$ removal, ammonia recycle,

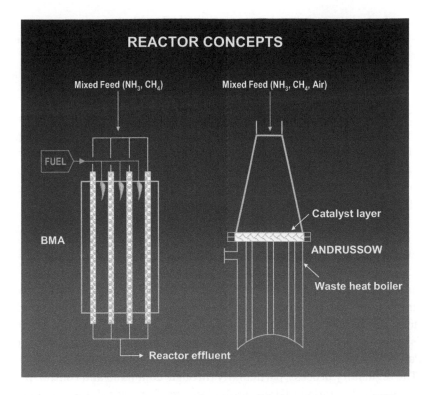

**Figure 15.3** Reactor concepts for BMA and Andrussow routes to HCN.

and energy recovery. At *macro* level there are potential synergies with gold producers in water and energy use, and in some locations with nickel refiners in hydrogen or ammonia use. Such process integration minimises resource consumption and waste formation, but implies increased capital and operational complexity.

Larger scale production tends to favour increased integration, but small-scale production at the point of use favours simple process designs and minimal capital.

An aqueous product is cheaper than a solid product in both capital and operating costs. If manufactured at site, costs and hazards of transport are eliminated. However, there are some disadvantages in making the aqueous product at its point of use:

- poorer scale economies result for a small plant sized for a local market

- higher financial risks are incurred should there be closure or production downturn at the gold mine; this risk could be reduced if a modular, transportable plant were practicable
- cost and hazards of ammonia (and to a lesser extent of caustic soda) transport to the manufacturing site are difficult to avoid.

The concept of manufacture at point of use remains an important principle in industrial ecology. While the right combination of site advantages and technology could conceivably work for sodium cyanide, there are clearly obstacles. The concept is more likely to be viable where the process is robust in terms of operating conditions, involves a single raw material and single product, and can be incorporated into a transportable plant.

## 15.9 Government Legislation

Regional and national legislation and international agreements have played an important part in driving improved environmental performance and cleaner production initiatives in industry. Some milestones in the US legislation and international agreements have been listed in the introduction to Part A of this book.

The regulatory process, particularly through 'command and control' policies, has some limitations. For example, damage often precedes legislation by a significant time interval. Further, legislation is costly and in some cases has been of limited effectiveness. Some countries such as Holland and Germany have taken leading initiatives in setting targets for emissions reduction and materials recycling. Many governments have attempted to encourage voluntary industry programs within their legislation. The Institution of Chemical Engineers through its journal 'Process Safety and Environmental Protection (Transactions Part B)' provides regular updates on UK and European environmental legislation.

Economic instruments can be used by governments to encourage greater environmental responsibility by individuals or companies. Examples of economic instruments include the following:

- emission charges related to the quantity and quality of pollutants emitted and the related damage
- user charges for collection, disposal, and treatment of wastes
- tradable or 'marketable' permits, which enable pollution (or resource consumption) control to be concentrated amongst those who can do it economically while capping total emissions. Trading

can be internal within a company for different company sites, as well as between companies. Tradable market permits have been used for $SO_2$ emissions in the United States in the 1990s and were introduced for $CO_2$ emissions in Europe in 2005 (European Union website) and New Zealand in 2010

- deposit refund systems involving a refundable deposit paid on potentially polluting products.

Economic policy instruments should be economically efficient, equitable, cost effective to administer, and successful in achieving improved environmental performance. Economic policy instruments and externality valuation are discussed by Turner, Pearce, and Bateman (1994).

## 15.10 Stakeholder Engagement

Stakeholder engagement is integral with the sustainability ethos, and is discussed in Chapters 1, 14, 16, and 17 in the context of sustainability concepts, sustainability assessment, and sustainable project development. Achieving effective stakeholder engagement and participation in process industry activities demands a thorough understanding of the interests, responsibilities, and the skills of the diverse groups that make up the stakeholders. This understanding requires ongoing consultation and education, best obtained by integrated planning through government, industry, professional societies, and educational institutions.

## 15.11 Lifestyle Implications

This book has focused on process engineering initiatives to achieve more sustainable process industries. The readiness of the chemical engineering profession to respond to the challenge of sustainability is encouraging, as instanced by the London and Melbourne communiqués. However, global problems such as large-scale poverty, continuing deterioration in climate impacts from global warming, and inadequate supply of raw materials, fresh water, and electricity in many parts of the word persist. There is also risk of technology solutions falling short in terms of timeliness and affordability. The extent of energy and fresh water consumption and waste emissions embodied in consumer goods, is seldom recognised or considered by consumers.

The potential should be explored for changing demand at the consumer level, particularly in more wealthy countries, by lifestyle changes in relation to

- reduced demand for consumer goods
- changing pattern of consumption of goods towards those which are less materials and energy intensive and have longer economic lives
- moderating excessive demands for personal living space and transport
- moderating excessive demands for heating and air conditioning in buildings.

Professional societies, industry bodies and academic institutions have an important role and responsibility in community consultation and education in the wider aspects of sustainability, including lifestyle implications.

## References

Benson, R. S. and Ponton, J. W. (1993) Process miniaturisation, *Trans. IChemE.*, 71A, 160.

Brennan, D. J. (1987) An alternative approach to modelling performance improvement in the process industries, *Chem. Engg. Austral.*, CE12(4), 7–12.

Brennan, D. (1988) Tracing S-curves for process technologies,' *Proc. Econ. Int.*, 7(2), 66–71.

Brennan, D. J. and Hoadley, A. F. A. (2003) *Chemicals Production at the point of use. The Case of Sodium Cyanide*, Paper 153 Chemeca, 9pp.

European Union website. http://ec.europa.eu/environment/climat/emission/ index_en.htm.

Hoffmann, V. H., Hungerbuhler, K., and McRae, G. J. (2001) Multiobjective screening and evaluation of chemical process technologies, *Ind. Eng. Chem. Res.*, 40, 4513–4524.

Kirk and Othmer (1994) *Encyclopaedia of Chemical Technology*, 4th edn, Wiley, New York.

Moens, L. (2006) Renewable feedsocks (Chapter 9), in *Sustainability Science and Engineering: Defining Principles* (ed Abraham, M. A.), Elsevier, Amsterdam, the Netherlands.

Turner, R. K., Pearce, D., and Bateman, I. (1994) *Environmental Economics*, Harvester Wheatsheaf, Hemel Hempstead, Hertfordshire, England.

Ullmann. (2002) Encyclopaedia of Industrial Chemistry, Wiley-VCH, Weinheim, Germany.

US Department of Energy. Energy Efficiency and Renewable Energy website: http://www1.eere.energy.gov/biomass/integrated_biorefineries.html

Van Beers, D., Corder, G., Bossilkov, A., and van Berkel, R. (2007) Industrial symbiosis in the Australian mineral industry. The cases of Kwinana and Gladstone. J. Ind Ecol, 11(1), 55–72.

Wadekar, V. V. (2000) Compact heat exchangers, *Chem. Engg. Prog.*, December, 39–49.

# Chapter 16

# Process Design and Project Development

## 16.1 Introduction

The provision of new process plants offers a special opportunity to link the best available technology with the objectives of sustainable processing. Design has an essential role in achieving this, whether applied to entirely new plants or to modifying existing plants. Chemical engineers make an important 'up-front' contribution in design by initiating and developing concepts of process schemes and the associated process plant.

Chemical engineering design interfaces with other engineering design, with expertise in plant construction, operation, and maintenance, and with specialists in criteria evaluation within the organisational framework of a project. In this chapter, we explore the design process, its role in process flow sheet development, and its context within the larger activity of project development.

## 16.2 The Design Process

The following characteristics may be attributed to design:

- Design involves the activities of synthesis and analysis:
  - ▶ Synthesis involves the creation and combination of elements to form a coherent whole.

*Sustainable Process Engineering: Concepts, Strategies, Evaluation, and Implementation*
David Brennan
Copyright © 2013 Pan Stanford Publishing Pte. Ltd.
ISBN 978-981-4316-78-1 (Hardcover), 978-981-4364-22-5 (eBook)
www.panstanford.com

> ▶ Analysis involves the assessment of the synthesised whole and the breaking down of the whole into elements for further study.

- Design occurs in response to a need which requires comprehension and definition.
- Design must satisfy certain criteria, most notably those derived from technical, economic, safety, environmental, social, and sustainability considerations.
- Design must be undertaken within constraints imposed by nature and by available technology, skills, money, and time.

The design process may be broadly depicted as shown in Fig. 16.1. Note that there are several feedback loops. Specifying the need in physical terms may lead to a closer examination of the need. Evaluation of an alternative may identify weaknesses which can be overcome by improved synthesis. Evaluation may reveal stringency in the defined need which can be partially relaxed. The design process depicted may be applied to a number of smaller elements of the design, with greater refinement of detail, until detail is adequate for the need to be realised as a concrete solution.

Design is the precursor to plant construction, commissioning, operation, and maintenance activities. As shown in Fig. 16.2, there are important feedback loops. The experience and expertise of construction, operating, and maintenance personnel is a vital input into the design of any future plant.

In chemical engineering design, five main activities can be identified.

- Process synthesis
  - ▶ synthesis and evaluation of alternative processes
  - ▶ recognition of technical, economic, safety and environmental constraints
  - ▶ evolutionary refinement of best option
  - ▶ use of modelling and optimisation techniques
- Process selection
  Applying suitable criteria for evaluation of established commercially available process routes to enable a balanced selection decision
- Process design
  Establishing the reaction conditions, separation processes, and mass, heat, and momentum transfer steps
- Equipment design
  Selection, sizing, detailed design, and specification of equipment
- Plant design
  - ▶ design of instrumentation and control systems
  - ▶ specification of utility requirements and utility systems

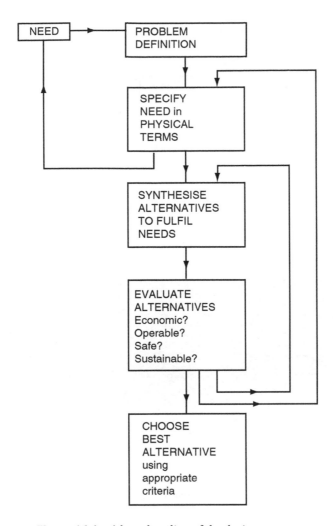

**Figure 16.1**   A broad outline of the design process.

▶ specification of effluents and requirements for their further treatment or disposal
▶ development of plant layout
▶ piping design including valve selection
▶ definition of operating procedures.

The activities in chemical engineering design cannot be rigidly segregated. For example, plant layout may influence the process flow sheet in terms of use of gravity flow or fluid pumping; available compression

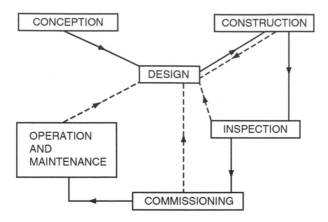

**Figure 16.2** Design within the wider project activity.

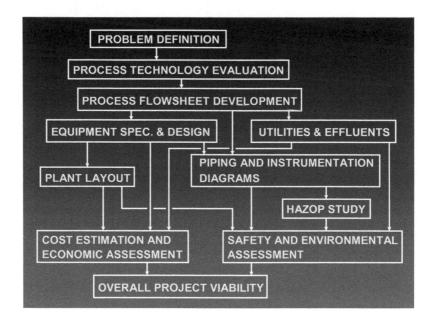

**Figure 16.3** Some key activities in chemical engineering design.

equipment may influence the pressure of gas streams. Equipment design constraints may influence the process flow sheet, and may require revision of the initial process synthesis. The five activities will normally involve more than one chemical engineer and involve liaison with other professional engineers. Many of the design decisions will be communicated to other professional engineers whose interpretations must be checked. In all activities, cost estimation, economic evaluation, and safety and environmental appraisals will be integral with the engineering work.

Figure 16.3 shows a typical sequence of chemical engineering design activities, and reflects the main flow of information through the overall design process.

## 16.3 Process Flow Sheet Development

We now review the evolution and development of process flow sheets, also referred to as process flow diagrams (PFDs), the criteria by which they are assessed, and guidelines for their documentation.

Five main steps can be identified, reflecting the activities in Fig. 16.1.

- Identify need and define problem.
- Specify need in engineering terms including feedstock, product specifications.
- Synthesise alternative processes to convert feedstock to product.
- Evaluate alternative processes using technical, economic, safety, environmental and sustainability criteria.
- Choose best alternative.

Some key aspects of process flow sheet development are now explored.

### 16.3.1 *Defining the Need*

Often the need is expressed in a way that requires increased definition. It may be a perceived market need, the need to treat an effluent from an existing plant, or the need to supply feedstock on a temporary basis while the normal feed source is unavailable. Ultimately, the need must be expressed as a process requirement in terms of raw material processed, the product produced, the required capacity of the associated process plant, and any other key requirements or constraints.

### 16.3.2 *Creating Plausible Solutions*

Various alternatives will usually be possible in terms of the process and equipment used to meet the need. It is important to create sufficient plausible solutions to ensure that the best solution is found. Process options matching possible process alternatives can be represented initially as simple block diagrams, where each block represents a stage in the process; these can be developed further, if appropriate, into simplified flow sheets.

### 16.3.3  *Screening of Alternatives*

A thorough evaluation and optimisation of each plausible solution would be a time consuming and expensive option. Hence, it is desirable to screen out less attractive options. The screening can be done using a range of criteria necessary to be met by the successful solution to the problem. The criteria include

- technical feasibility
- capital and operating cost
- safety
- environmental acceptability
- sustainability
- reliability and operability.

Some of the above criteria might conflict. Some options may be expensive in capital costs but low in operating costs. Other options might be costly in meeting regulatory environmental standards.

### 16.3.4  *Further Evaluation of Selected Options*

Desirably the screening will limit the options to a small number, say three, which can then be more thoroughly evaluated. When the best solution option is chosen, some optimisation will still be needed in terms of efficiency in energy and raw materials usage, minimising effluents, and trade-offs between capital and operating cost. Process simulation packages are a valuable aid both to optimisation and to exploring effects of deviations from base assumptions, such as feedstock and product compositions. It must be recognised that there are some aspects such as ease of operation and reliability where a process simulation package has limited use; for these criteria, other data sources and (particularly) operational experience must be considered.

### 16.3.5  *Optimisation and Scrutiny of Final Solution*

Ultimately, the chosen solution must be optimised and pass acceptability tests in the various criteria. While optimisation methods and computer packages are valuable tools, the optimum solution reflects the input assumptions made. Since some deviations from input assumptions are inevitable in plant operations, it is important to explore how robust the design is to such deviations. One important aspect is the influence of deviations on quantities and compositions of effluents. The final solution should also be scrutinised to ensure that it meets all the criteria of acceptance. These criteria also have an essential role as inputs into the early stages of process conception and development.

## 16.4  Criteria for Process Flow Sheet Evaluation

Evaluation criteria are now discussed under their various categories.

### 16.4.1  *Technical Feasibility*

Meeting all the fundamental technical requirements, for example,

- meeting laws of chemistry, physics and thermodynamics
- constraints of mass and energy balances
- temperature and concentration driving forces.

### 16.4.2  *Capital Cost*

Factors contributing to capital cost include

- plant complexity (including numbers of process steps, extent of recycles)
- design pressure and temperature
- materials of construction
- plant capacity
- extent of storages, utility generation and buildings
- stringent safety and environmental standards.

### 16.4.3  *Operating Costs*

Factors contributing to operating cost include

- raw materials consumption and unit cost of raw materials
- utilities (or energy) consumption and unit cost of utilities
- monitoring, treatment and/or disposal of effluents
- personnel employed, salaries and payroll overheads

- capital related costs including depreciation, maintenance, and insurance.

### 16.4.4 *Safety*

Factors contributing to process safety include

- properties of process materials, from feedstock to product
  - ▶ toxicity
  - ▶ flammability and potential for fire and explosion
  - ▶ corrosiveness
- inventories of hazardous materials
- extremes of pressure, temperature
- development of overpressure or underpressure.

### 16.4.5 *Environmental*

Major contributions to environmental damage are

- waste emissions, leading to range of environmental impact categories
- resource depletion through excessive use of raw materials and utilities.

### 16.4.6 *Sustainability*

Sustainability implies the capability to meet and sustain standards in economic, environmental, safety, and social criteria over plant life. Considerations include

- assured and adequate cash flows over project life
- assured markets for products and co-products
- security of supply, quality, and price of raw materials
- feasibility of using renewable sources of raw materials and energy
- sustainability merits of product manufacture, use and disposal
- implications of any product packaging or distribution for environment
- long-term effects on health of employees, site, natural environment.

### 16.4.7 *Reliability*

Reliability is important in terms of

- meeting required product quality
- operating to the required performance standards matching the criteria
- achieving required production capacity, both in terms of maximum production rate in tonnes per day, and operating days per year.

Drivers for reliability include

- interdependency of distinct process plants
- customer demands in terms of product supply continuity
- capital intensiveness of plants and resultant cost of downtime and underused capacity
- prolonged start-up and shut-down routines
- minimal storage capacity for raw materials and products.

Means of achieving reliability include

- avoiding unnecessary complexity in the process
- robustness of plant and equipment to changes in operating conditions
- selecting correct materials of construction
- providing equipment redundancy, for example, pumps in parallel, multiple reactors in parallel
- reliable utilities.

### 16.4.8 *Operability*

Ease of operation, control, and maintenance.

## 16.5 Process Flow Sheet Documentation

As is evident in Fig. 16.3, process flow sheet development is a key design activity influencing the outcomes of many downstream design activities. The process flow sheet is arguably the most consulted of all project documents, with consultation extending beyond the design phase of a plant well into its operating life. Key aspects of complete and clear documentation of the process flow sheet are outlined in Appendix 1. Some examples of published process flow sheet documents with material flows include

- Figure 1.1 (Benzene manufacturing plant) in Wells, Seagrave, and Whiteway (1976)
- Figure 4.2 (Nitric acid plant) in Sinnott and Towler (2009).

Flow diagrams for hydrocarbon processing are provided in Turton, Bailie, Whiting, and Shaeiwitz (1998).

Some important outcomes from the process flow sheet which influence environmental performance include

- raw materials consumption
- utilities consumption
- process effluent streams.

These outcomes should be separately documented.

Process flow sheets are to be distinguished from other important design flow sheets and diagrams, namely,

- *piping and instrumentation diagrams*, whose function is to show all piping, valves, instruments, control loops.
- *simulation flow sheets*, which are intended to illustrate a computer simulation. These diagrams, for example, show stream junctions as 'mixers' which do not represent equipment items.
- *layout diagrams*, which are intended to show plan and elevation views of equipment within the plant.

## 16.6 Piping and Instrumentation Diagram

The piping and instrumentation diagram (P&I diagram) is developed from the process flow sheet and design and specification of equipment to a point where the piping system for the plant as well as instrumentation and process control system are defined. A number of P&I diagrams will be developed for a process plant detailing piping, valves, fittings, and instrumentation.

The P&I diagram is also a key source of information for identifying potential sources of fugitive emissions, provisions for drainage, venting, purging, and the maintenance and cleaning of equipment and piping. As such, it is a key document for identifying potential waste sources in operations, such as fugitive emissions and accidental releases or spillages, which were discussed in Chapter 9. The P&I diagram is also an essential document for conducting a HAZOP study in safety assessment. Examples of P&I Diagrams can be found in the references listed above under section 16.5.

## 16.7 Project Development

The design process takes place within the organisational framework of a project, where key personnel, skills, and resources are brought together to develop and implement the proposed scheme. Projects offer exciting opportunities to deliver major economic, environmental, and social benefits within industries and at national and international levels, and to serve the common good. At the same time, failed projects can leave a trail of devastation in their wake in terms of wasted capital, loss of human life, environmental damage, ongoing litigation, bad will between project participants, poor profitability, and so on. Many important elements are required to make a project successful; neglect or failure in any one of these elements can be sufficient to ruin the project.

Projects vary enormously in concept and scope, in level of capital expenditure, and in time schedules for development and implementation. Small projects, while not involving large capital sums or personnel requirements, still have the potential to make important improvements if successful, and to cause significant damage if unsuccessful. *It is a mistake to resource a small project with inadequate capital, inadequate time for planning and execution, or inadequately skilled and experienced personnel just because it is small relative to a company's portfolio of projects.*

Project development is the formative stage of a project prior to *sanction*. At sanction, the project and its related capital allocation are approved by the board of directors of the company which will later operate the related process plant. Some key steps prior to sanction are shown in Fig. 16.4.

The time span from identifying an investment opportunity to sanction varies considerably; 2 years would not be unusual where the technology is established, but longer periods are often required, especially if further research and development is necessary. The investment opportunity is

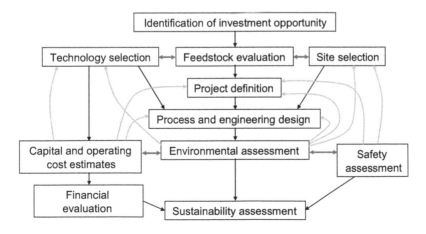

**Figure 16.4** Key steps prior to project sanction.

characteristically a market for products, but may alternatively be driven by cost saving or environmental compliance.

Following sanction, more detailed design, equipment and materials procurement, and plant construction take place. When construction is complete, the plant is commissioned for operation. Time spans from project approval to commissioning vary considerably depending on project size and complexity; the time span could range from 1 year for a small chemical plant to 3 years for a world scale ethylene plant to 5 years for a large coal fired power station.

At commissioning, the production capacity and performance capability of the plant is proven. After commissioning, the operating life of the plant commences and continues until a decision is made to terminate operation. At termination of operations, the plant is decommissioned and either 'mothballed' or dismantled. Typical operating lives for process plants assumed at the investment phase are 10 to 15 years, but many plants have much longer lives.

An important aspect of successful design is an effective design — operator interface (Brennan, 2006) encompassing

- clear understanding by designers of the nature of plant operations and the roles of operations personnel
- clear understanding by designers of the drivers acting on operations
- clear communication of design intent to operations personnel
- recognition of the learning process in design and operations
- recognition by designers that plant operators and maintenance workers, as well as plant components, are failure prone
- interaction between design and operations personnel in the development and review of design work.

## 16.8  Acceptability Criteria for Projects

Acceptability criteria for projects mirror those for process designs, and undergo scrutiny from a larger group of stakeholders, particularly those with responsibilities for managing, approving and funding the project.

### 16.8.1  *Technical Requirements*

A basic requirement for process plants is that they work. If a number of similar plants have been built and operated previously, the risks are greatly reduced although there may still be changes from previous plants in terms of feedstock composition, equipment choices, and design innovations.

These changes may be constrained by circumstances or may offer perceived benefits, but can also be sources of risk in achieving the required plant capacity and performance.

Greater risks attend the scaling up of new processes demonstrated at pilot scale, or adapting existing plant designs to situations where there are significant changes (e.g., in mineral feedstock compositions for mineral processing plants). The commercial pressure to seize a market opportunity can conflict with the need for due diligence in process and plant development work to minimise technical risk. Scaling up is inherently risky, since it is difficult to maintain the same chemical, thermal and mechanical performance parameters from one scale to another. Scaling up a stirred vessel in which mixing, heat transfer, solids suspension, and chemical reaction are all important, is a good example. Entire books have been written on scale up, but it remains a major challenge in process development. Many projects have failed because of excessively ambitious leaps in scale.

Selection of construction materials for plant and equipment is a key decision and can be risky for new processes, or when using different feedstocks from those of previous plants (Fontana, 1986). Effects of temperature, velocity, and impurities can be complex, for example, when considering metals and alloys in contact with inorganic acids. Velocity and turbulence effects can cause erosion corrosion, while chlorides can cause pitting or stress corrosion of stainless steels. Corrosion testing prior to design can be useful if done competently, but there may still be problems which lay hidden until the plant is operating.

Often, projects may involve not just one plant, but several interdependent process plants. The complexities associated with successfully bringing all plants on stream simultaneously at commissioning to achieve required capacity and performance can be considerable, particularly where there are interconnecting process or utility streams between the plants. Phased commissioning of the individual plants is normally a wiser strategy.

## 16.8.2 *Economic Viability*

The economic viability of a project is of great importance, and there are many potential areas for failure. Factors shaping the economics are shown in Fig. 16.5.

Markets are of key importance in ensuring profitability. Product selling price is the most sensitive parameter governing profitability, while plant capacity must be matched to market volume. Markets will drive the potential for economic success of the project over its life, so market forecasting

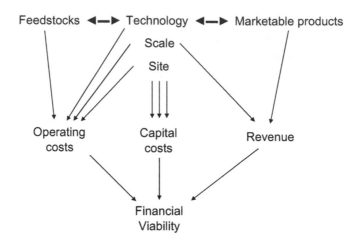

**Figure 16.5**   Some factors shaping the economics of a project.

over the horizon of construction and operational phases is essential. While forecasting is inherently uncertain, various scenarios should be systematically examined. These include analysis of competing products, processes and businesses, prospects of surplus production capacity, and external factors such as currency exchange rates.

Feedstocks must be matched to the technology and product, and feedstock supply must be secure over project life in terms of availability, price, and quality. Where alternative sources or compositions of feedstock are feasible over project life, the robustness of plant design to cope with these changes should be examined.

Inside battery limits capital will be influenced by technology, plant capacity, and project management, and also by the plant site and the investment climate at the time of project implementation. Total capital incorporating outside battery limits investment and working capital will be influenced by the site and logistics.

The concept of life cycle costing is commonly acknowledged as a precursor for economic success of a project; as part of this, it is important to weigh up incremental capital expenditure with corresponding savings in operating cost. However, there are divergent interests of different parties involved in project development and execution. For example, contractors must deliver a working plant on time, within budget and within performance specification, but the operating company must bear the capital investment incurred and operate, maintain, and develop the plant over a much longer time horizon. Such divergent interests must be recognised and resolved to achieve successful project outcomes. Penalties of over expenditure and excessive lead times are obvious, but penalties arising from too lean a capital budget can also be serious and can limit plant performance in its operating life.

Operating costs are influenced by site location, the process technology adopted and the quality of design and construction, as well as the calibre of operating personnel employed.

In exploring profitability based on cash flow projections, an important aspect is the possible time dependence of economic parameters over project life (Brennan, 1998). For example, plant capacity will be influenced by ease of 'ramping up' production in early operating life, as well as the extent of learning, exploitation of capacity margins, and new technology introduction in later operating life. Product selling price (in real terms) may decline under the influence of competition from improved products or processes during operating life. Plant operating life itself is always an uncertain parameter.

Thus, it is clear that a thorough analysis of profitability and attendant risks should be made during the project planning and development phase; this analysis should influence key decisions regarding all elements of the project. Competitive analysis of operating costs (proposed project vs existing operating plants) should always be used as a parallel approach to profitability projections based on cash flow analysis. Vulnerability to economic failure in upstream and downstream links of the product life cycle must also be considered.

## 16.8.3 *Safety*

Techniques for hazard identification and risk assessment are well known and established as an integral part of project development (Crawley & Tyler, 2003). Systematic approaches to inherent safety are less widely practised, even though the principles have been expounded for at least two decades (Khan and Amyotte, 2003). The principles of inherent safety have been summarised in Appendix 2 of Chapter 12. The maxim of changes being least difficult and costly in the earlier phases of project development implies that early use of inherent safety methodology and related quantitative assessments should be mandatory.

AIChE (1989) argued for safety review procedures at five stages of a project from project inception, engineering, detailed engineering, procurement and construction, and commissioning. AIChE (1989) also provides guidance in site selection, plot plan development, and risk screening in technology choices as part of the inception phase.

According to Crawley (2004), there have traditionally been six stages of process hazard analysis in the life of a project, to which Crawley and Tyler (2003) have recommended the addition of two further stages for *inherent safety* and *abandonment/demolition*:

- *inherent safety*
- concept selection
- front end engineering development design ('FEED') sometimes called concept development
- detailed design
- construction
- commissioning and start-up
- post start-up review
- *abandonment/demolition*.

Newer technologies may need more careful risk assessments than established technologies where many of the risks have become evident through the benefit of experience.

### 16.8.4 *Environmental*

The cleaner production philosophy and life cycle assessment methodology discussed in earlier chapters are particularly relevant and applicable to the development phase of a project in terms of

- defining the product life cycle, and identifying key environmental impacts at the various stages of the life cycle;
- defining system boundaries relevant to the project, and their relationships with other interdependent systems;
- adopting the preventive approach to both waste emissions and inefficient resource consumption;
- adopting a holistic view of the full spectrum of environmental impacts arising from the project.

These aspects are important for an emerging project with its choices of site location, technology selection, raw materials and utilities sourcing, effluent management, and product packaging and distribution. The aspects are also important in terms of the competitive position of the project relative to other operations in the same industrial sector.

## 16.8.5 *Sustainability*

As discussed in Chapter 1, the principles of sustainability imply longer time frameworks in project evaluation than those traditionally used. Extended time frameworks originate from the environmental considerations of resource depletion and impacts from emissions, and are also pertinent to the social consequences of a project, the long-term potential for a product or process, and the future use of a plant site. Rather than considering the proposed plant solely over its operating life, the implications of its replacement or next generation plant should also be considered, as well as the long term consequences of any net waste remaining after treatment, storage and disposal of effluents.

Sustainability also implies proper consideration of the implications of the project for the various stakeholders, for example, the need to satisfy their entitlements, to brief them, and consult with them when appropriate. Relevant stakeholders include employees, the resident community nearby, the wider community, government, customers, and shareholders. In terms of the wider community, openness, transparency, and the appropriate level of detail, as well as timing, are important considerations. Some stakeholders have the potential to provide valuable input into project conception and execution.

Azapagic (2003) identified a list of typical stakeholders for projects, with differing levels of their concern regarding economic, environmental, and social implications, and differing time scales relevant to these concerns. The approach outlined by Azapagic provides a useful first step to developing a more expansive approach to identifying and analysing stakeholder concerns and impacts, and to developing strategies for addressing them.

There are potential conflicts of interest between various stakeholder groups providing obstacles to achieving satisfaction for all groups. For example, some industry projects may offer secure employment to employees but may be high $CO_2$ emitters with adverse implications for the global community. The effective integration of stakeholder concerns to achieve a desirable project outcome is an even more demanding challenge than effective consultation with stakeholders.

Increasingly, within more complex projects, alliances are sought between interested parties where projects cross boundaries between separate business organisations or entities. Such bodies might be infrastructure suppliers (port and jetty), utility suppliers, effluent processors, feedstock refiners, and so on. Good long-term financial, safety, and environmental outcomes for relevant partners is important to ensure overall project success.

## 16.9 Integrating Criteria Assessments

Much diverse expertise is involved in performing, interpreting, and applying the technical, economic, safety, environmental, and sustainability assessments in projects. The drawing together of this expertise (Fig. 16.6) and the related specialist skills to achieve the best project outcome is a formidable challenge. Not only are economic, safety, and environmental assessments made by different personnel with distinct experience and skills, but even within economic criteria, market, capital cost, and operating cost assessments are normally made by different personnel. Capital cost estimates rely on design, procurement, and construction expertise. Estimates of personnel and maintenance requirements within operating cost estimates rely on plant operating experience. Market volume and price estimates draw on knowledge of products, markets, business cycles, and competitor activities.

Trade-offs within and across criteria (as discussed in Chapter 14) must be considered and resolved. Not all of the project management decisions will correspond to 'win-win' outcomes. Figure 16.4 depicted a view of the steps involved in a process industry project prior to sanction. These steps are often thought of as a linear flow of information and decision making. What is required in practice, however, is a more concurrent approach to the various activities, including the criteria assessments. There is need to feed insights gained by the criteria assessors progressively back into the key decisions of technology selection, site selection, feedstock selection, project

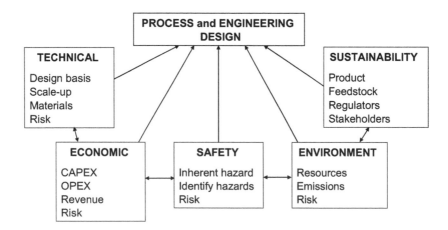

**Figure 16.6** Criteria assessments of process and engineering designs.

| | Capital cost | Operating cost | Safety | Environmental |
|---|---|---|---|---|
| 1 | Factored cost estimate from purchased equipment cost | Feedstock, utilities, factored fixed costs | Inherent safety assessment and hazard identification | Life cycle assessment over bounded system |
| 2 | Detailed estimate of all materials and labour costs | Detailed estimate of all variable and fixed costs | HAZOP studies: layers of protection analysis | Site specific environmental impact assessment |

**Figure 16.7** Staged reviews and assessments during project development. Stage 1 represents an early stage, and Stage 2 a later more detailed stage.

definition and scope, and design of process and plant. This more concurrent and interactive approach contrasts with an approach which relies solely on approval or amendment decisions at various stages of project development. It has the benefit of ensuring that project development time is shortened, and is consistent with the accepted maxim that changes are least difficult and costly in the earlier phases of projects.

Selecting the appropriate assessment method at different stages of project development is a further challenge. Inherent safety and life cycle assessment methodologies are particularly useful at the inception phase, but capital and operating cost estimation methods also need tailoring. Use of staged reviews is illustrated in Fig. 16.7.

How do we ensure that the various specialist groups exploring and applying assessment criteria have adequate interaction to ensure the best outcome for a project? How can we ensure that specialists in one area have adequate appreciation of other specialists' skills, priorities, and perspectives? And how can we ensure that the project manager has adequate understanding of the principles of the various specialisations and the implications of their findings? Two of the important principles are

1. to bring together the key parties contributing to the design and its assessment at the earliest possible phase of the project, to agree project objectives, and plans for implementation
2. to establish mechanisms for regular consultation and progress reviews by these key parties.

One widely adopted approach has been to group health, safety, and the environment into one area for assessment. Trade-offs are often still required, however, for example, when one design option offers improved safety but higher environmental burden compared with an alternative design. Crawley (2004) has suggested the possibility of assigning money values to potential loss of life, loss of a material resource, and impacts from emissions to help resolve such conflicts. This approach has been discussed in Chapter 14 in the context of estimating environmental externalities, but the uncertainty and subjectivity of assigning money values, especially in relation to human life, remains a difficulty.

## 16.10 Concluding Remarks

There are many dimensions to the successful design, development, and management of process industry projects. It is important for chemical engineers involved in design and project activities to understand their role in the wider project activity.

For projects to be successful, threshold standards must be met in technical, economic, safety, environmental, and sustainability criteria. These criteria must influence a spectrum of decisions including site location, technology choice, scale of operation, feedstock source, investment timing, and process and plant design. Project proponents must engage with stakeholders including governments and the wider community to ensure project success.

It is a demanding task for the project manager to ensure that criteria assessments are made competently, that adequate information is exchanged between specialist functions, and that the various assessments effectively shape project development. Management tools for supporting such concurrent and interactive work among specialists are a priority. Establishing early liaison and interaction is an important first step.

Commitment to satisfactory performance over the entire life cycle of the project is an essential element of project conception and development. Plant operating life will almost always be the longest part of the life cycle. It should be acknowledged that there is some ambiguity surrounding the exact terminal point of the project life cycle. Even after plant closure, there may be residual effects on site condition, accumulated technological and business know-how, and stakeholder relationships.

Advances and refinements in evaluation methodology and techniques, particularly in environmental and safety criteria are continually being made. Such advances should be incorporated into project decision making and have a role in communicating project merits to stakeholders.

# References

AIChE (1989) *Guidelines for Technical Management of Chemical Process Safety*, AIChE, New York.

Azapagic, A. (2003) Systems approach to corporate sustainability. A general management framework, *Trans. IChemE.*, **81**(B), 303–316.

Brennan, D. (1998) *Process Industry Economics*, Institution of Chemical Engineers, Rugby, England.

Brennan, D. (2006) Design – operation interface in chemical engineering, *Trans. IChemE.*, Part D, **1**, 95–100.

Crawley, F. K. (2004) Optimising the life cycle safety, health and environmental impact of new projects, *Trans. IChemE.*, Part B, **82**, 438–445.

Crawley, F. K. and Tyler, B. (2003) *Hazard Identification Methods*, European Safety Center, IChemE, Rugby, England.

Fontana, M. G. (1986) *Corrossion Engineering*, 3rd edn, McGraw-Hill, New York.

Khan, F. and Amyotte, P. (2003) How to make inherent safety a reality, *Can. J. Chem. Engg.*, **81**, 2–16.

Rudd, and Watson, (1968) *Strategy of Process Engineering*, Wiley, New York.

Sinnott, R. K. and Towler, G., (2009) *Chemical Engineering Design*, vol. 6, Elsevier, Oxford, England.

Turton, R., Bailie, R. C., Whiting, W. B., and Shaeiwitz, J. A. (1998) *Analysis, Synthesis and Design of Chemical Processes*, Prentice Hall, Upper Saddle River, New Jersey.

Wells, G. L., Seagrave, C. J., and Whiteway, R. M. C. (1976) *Flowsheeting for Safety*, Institution of Chemical Engineers, Rugby, England.

# Appendix 1. Documentation of Process Flow Sheet

## A. Purpose of Process Flow Sheet

1. To convey the details of the process clearly and concisely
2. To document key information about the process and assist verification of mass and energy balances
3. To provide a single reference point for identifying all major equipment items in the plant and their role in the process
4. To enable easy cross-reference in a design report from

   (a) process description
   (b) presentation of mass and energy balances
   (c) equipment specification
   (d) equipment costs within capital cost estimate

## B. Essential Parts of the Process Flow Sheet

1. Flow diagram showing equipment items and process and utility streams
2. An equipment list table
3. A table detailing process and utility streams
4. A title block

## C. Flow Sheet Representation

1. Raw materials should enter on the left and product streams leave on the right of the diagram.
2. Key terminal raw materials and products streams, as well as any process effluent streams, should be clearly identified on the diagram.
3. Equipment items should be *roughly* scaled in proportion to size and *roughly* placed at elevation reflecting elevations on plant site.
4. Direction of flow should be clearly indicated by *arrows* on process streams.
5. Process streams should be systematically numbered.
6. Where possible the diagram should be on one piece of paper. For complex plants, multiple diagrams may be necessary with clear linking of streams.
7. Use clear, uncluttered diagrams with tabular detailing of process streams and equipment.
8. Provide a title block, defining the author, author's affiliation, the process, and the flow sheet capacity (e.g., tonnes of product per day).

## D. Process Stream Tables

1. For each process stream, detail

   - identifying number
   - mass flow rate
   - temperature
   - pressure
   - phase
   - composition.

2. For each utility stream, detail key conditions, for example, temperature, pressure.

## E. Equipment List

Identify each equipment item with letter prefix, number, and title defining equipment function.

## 1. Examples of letter prefixes

**E** Heat exchangers    **V** Vessels    **R** Reactors
**C** Compressors         **P** Pumps     **S** Storage tanks
**T** Towers                **F** Filters

## 2. Numbering

E1, E2, E3, E4, and so on.

## 3. Titles

These should be concise but also definitive, for example,

C1 — Hydrogen feed compressor
C2 — Hydrogen recycle compressor
P1 — Strong acid circulating pump.

## F. Graphical Symbols for Equipment Items

Examples of symbols for different equipment items are provided in Appendix A in Sinnott and Towler (2009). Ensure the symbol is as good a representation as possible of the type of equipment envisaged.

# Chapter 17

# Operations Management

## 17.1 Operational Phase of a Process Plant

The operating life of a process plant is characteristically by far the longest part of its life cycle. It is in operations where the benefits of technology development, design, and capital investment are realised. It is here where income generated rewards capital expenditure, where further employment is created, and where products fulfil market demands. While the manufacture of plant construction materials contributes to environmental impacts, by far the greatest impacts during project life normally occur in operations, resulting from consumptions and emissions within process and utility streams. In plant operations, waste emissions can occur beyond those identifiable from design intent or process flow sheet analysis, in ways identified in Chapter 9.

In considering the operational phase of a plant, it is important to understand the nature of plant operations, the personnel responsible (from plant operator through to management), and the many drivers acting on operations. These drivers include

- product quality demands
- production targets for quantities of product produced within time frameworks
- the reliability and availability of plant to ensure continuity of supply and achievement of production targets
- performance targets such as raw materials and utilities consumptions per tonne of product

*Sustainable Process Engineering: Concepts, Strategies, Evaluation, and Implementation*
David Brennan
Copyright © 2013 Pan Stanford Publishing Pte. Ltd.
ISBN 978-981-4316-78-1 (Hardcover), 978-981-4364-22-5 (eBook)
www.panstanford.com

- standards of safety, both occupational and public
- environmental standards and their implications for allowable emissions
- maintenance requirements, including inspection and monitoring of plant and equipment, as well as budgetary constraints and shutdown planning.

These drivers have key influences on the economic, safety, and environmental performance of the plant and hence its sustainability.

In terms of environmental standards, emissions must comply with regulatory license requirements, but should also be acceptable to local communities, where noise and odour issues often become problematic.

While plants are designed for specific process conditions, and specific feedstock and product specifications, events change over plant life. Feedstock compositions may be variable, product specifications may become tighter and environmental legislation may become more stringent. Market growth may stimulate deliberate expansion of production capacity. Plant modifications are frequently required, driven by technology improvements, opportunities for cost reduction, the need to address operating and maintenance problems, and the need for improved environmental and safety performance. Such modifications can contribute to improved capacity and performance over operating life.

The importance of an effective design – operations interface has been discussed in Chapter 16. Just as it is important for designers to make clear their design intent for the benefit of stakeholders including operations personnel, it is also important for operations personnel to seek to understand the design intent behind the processes and plants they operate. This is true both for new plants, and also for smaller projects involving plant modifications during operating life.

## 17.2  Plant Maintenance

Sustainable operations depend heavily on the quality of maintenance on equipment, piping and valves, instrumentation, and the entire plant. Maintenance is required on both a planned basis (usually during an extended plant shut-down) and on an unplanned basis, where it is necessary to respond to unforeseen faults.

The penalties for inadequate maintenance standards include failure of plant components leading to a wide range of deviations from intended

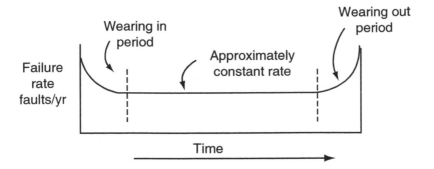

**Figure 17.1** 'Bath-tub' curve showing failure rates of plant components over time.

performance, possible releases of process materials with safety and environmental consequences, and forced shut-downs. Forced shut-downs imply interrupted production, reduced plant availability, loss of sales revenue, and unforeseen costs of repair.

Routine inspection and monitoring of a process plant is one essential aspect of preventative maintenance. There are aids to this, such as non-destructive testing including ultrasonic and acoustic techniques, and analytical sampling devices for detecting leaks of flammable or toxic fluids. However, regular and systematic on-site inspection of plant is an essential part of effective maintenance.

Failure of plant components has often been modelled by the so-called 'bath-tub' curve (Fig. 17.1). The bath tub curve shows three stages:

- an initial (or 'wearing in') period where failure rate decreases due to an adaptation and learning process;
- a prolonged period of approximately constant failure rate;
- a final period where failure rate increases due to wear or corrosion of components.

Some failure rate data have been provided for equipment items by Turner (1977) and for instrumentation and control components by Lees (1986).

Maintenance of equipment, instruments, piping, and valves requires shut-down of the related section of plant, any necessary drainage, venting and purging, and isolation from process streams, electricity, and utilities. Provision must be made at the design stage of plants for maintenance activities encompassing these features. This is addressed during the development of the piping and instrumentation diagrams for the plant. Strict procedures are required in operations to ensure the plant is safe for maintenance activities to proceed. These procedures are referred to as 'permit to work' systems. Special procedures are required for work within confined spaces and closed vessels.

Details of maintenance requirements in varying contexts of a process plant's life cycle are provided by Townsend (1992) and Scott and Crawley (1992). There is an important need for an active interface between design and maintenance personnel during the design stage of plant, and between operational and maintenance personnel during the operating life of a plant.

## 17.3 Environment Management Systems

To achieve environmental objectives, environmental management becomes mandatory for achieving sustainable outcomes in the operating life of a plant. Environmental management systems (EMS) are covered by the International Standard ISO 14001. The international standard proposes the steps identified in Fig. 17.2 and amplified in the following text.

### 17.3.1 *Commitment and Policy*

An organisation should define its **environmental policy** and ensure commitment within the organisation to its EMS. Most industrial corporations have had a documented environmental policy for some years.

### 17.3.2 *Planning*

An organisation should define its environmental plan to fulfil its environmental policy. Aspects of planning include

- identification of environmental aspects and evaluation of associated environmental impacts
- legal requirements
- environmental policy
- internal performance criteria
- environmental objectives and targets
- environmental plans and management program.

### 17.3.3 *Implementation*

An organisation should develop the capabilities and support mechanisms necessary to achieve its environmental policy, objectives, and targets. Important aspects are as follows:

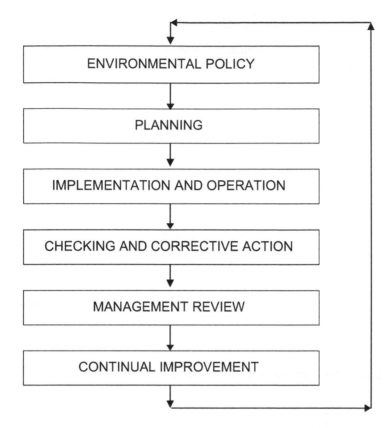

**Figure 17.2** Outline of steps for an environmental management system.

- Capabilities include human, physical (plant and equipment), and financial.
- EMS elements should be integrated with existing business management.
- Responsibility for the effectiveness of the EMS should be assigned to an appropriately skilled senior person.
- Senior management must ensure employees are aware and motivated. This implies adequate knowledge, skills, and training.
- Support action includes adequate communication and reporting of environmental activities, as well as systematic documentation of operational procedures. Suitable controls must be implemented and emergency plans and procedures established.

### 17.3.4 *Measurement and Evaluation*

An organisation should measure, monitor, and evaluate its environmental performance by

- taking corrective and preventive action
- maintaining EMS related records
- conducting audits of the EMS on a regular basis.

### 17.3.5 *Review and Continuous Improvement*

An organisation should review and continually improve its EMS with the objective of improving its overall environmental performance. The review should be broad enough in scope to address the environmental dimensions of all activities, products, and services of the organisation, including their impact on financial performance and competitive position. A process of continuous improvement should be adopted. Performance should be compared with policies, objectives, and targets to identify improvement opportunities.

## 17.4 Environment Improvement Plan

Environment Improvement Plans (EIPs) have been required in Australia in some states under state Environment Protection Authority (EPA) licence conditions. EIPs are also voluntary initiatives by some companies in other Australian states. EIPs involve a public commitment by a company to enhance its environmental performance. A number of companies have drawn up EIPs as voluntary commitments after consultation with their local communities.

Examples of EIPs in Victoria, Australia, include those of Qenos Australia and Alcoa.

Under such EIPs, companies can obtain accredited licenses which confer benefits of simplified regulation. In return companies implement an EMS and an EIP which are acceptable to the Victorian EPA and to the community, and which undergo regular audits by EPA-approved auditors. An EIP should include

- a requirement that any relevant state policy, regulation or licence condition must be complied with
- emission and waste production standards for the industry
- requirements for the monitoring of compliance with the EIP

- provision for community participation in performance evaluation in meeting objectives of the EIP
- provision for upgrading of plant to meet EIP objectives
- provision for assessment of new or emerging technology
- provision for contingency or emergency plans.

## 17.4.1 *Examples of EIPs*

Qenos Australia P/L has been awarded accredited licensee status under the Victorian EPA. The EIP is a public document. Action plans in Qenos' 2009 EIP included steps relating to the following aspects of environmental performance:

- greenhouse gases
- odour
- transported Wastes
- water and Wastewater
- soil and Groundwater
- flares
- training.

Copies of EIPs can be accessed through company websites. At the time of writing, copies of the Qenos and Alcoa EIP documents can be accessed through http://www.qenos.com.au/internet/home.nsf/vwfiles/environ-mental_action_plan_2009/ and http://www.alcoa.com/australia/en/info_page/EIP.asp, respectively.

## 17.5 **Responsible Care**

Responsible Care is a voluntary chemical industry program to manage health, safety, and environment, initiated by the chemical industry in Canada in 1983. The program was modified and adopted by US Chemical Manufacturers Association in 1988. In 1989, a Responsible Care program was adopted by the chemical industry in Australia, and later in several European countries. The Australian program is administered by Plastics and Chemicals Industry Association (PACIA). The Responsible Care program has five key elements:

- principles which recognise public concerns and industry's desire for self-improvement
- codes of management practice

- national Community Advisory Panel
- member self-evaluation of performance
- executive leadership groups.

Further information can be accessed through http://www.responsible-care.org.

## 17.6 Environmental Performance Monitoring

Reliable data for resource consumptions and emissions demand a systematic monitoring program. Emissions monitoring in the workplace is important from a number of perspectives.

Operator health and safety can be threatened by continuous exposure to toxic materials as the result of regular duties in the workplace and by short-term exposure from abnormal releases. In this context, permissible concentrations in the workplace have been defined in terms of threshold limit values (TLVs). TLV data are available both on a time-weighted average basis (TLV-TWA) assuming normal working conditions of 8 hours per day 5 days per week, and permissible short-term exposure limits (TLV-STEL) based on an exposure of up to 15 minutes.

Continuous sampling of air at strategic points around a process plant is necessary to detect leaks of process materials which threaten operator safety. One example is the continuous monitoring of air around a PVC manufacturing plant in order to detect any vinyl chloride emissions.

Regular monitoring of air quality at the plant site and within site boundaries is important to ensure minimum quality standards are met from an environmental viewpoint. This is an important part of fulfilling license agreements with environmental regulatory bodies. Planned monitoring of releases at potential fugitive emission sources is also an important part of plant control.

Environmental performance monitoring is also required at plant boundaries to ensure that neighbouring communities are not exposed to dangerous or offensive materials. Little is published regarding acceptable concentrations of chemicals for continuous exposure by the public. One suggestion has been to use 2.5% of the TLV-TWA. One frequent aspect of concern at plant boundaries and in neighbouring communities is odour. Odorous materials are not always identified at the design stage of a new plant or when planning modifications to an existing plant. Perception of odours and response to odours varies between individuals, but persistent and intense odours are often the basis for complaint (Heinsohn & Kabel, 1999). An interesting case history of an odour problem arising from an alumina refinery is discussed by Coffey and Donoghue (2006); the case study is of interest because of the technical, operational, environmental, and community contexts discussed.

In predicting and assessing airborne concentrations of chemicals and particulates following release from a point source, dispersion modelling is required. The degree of dispersion is affected by the nature and location of the release within the plant, topography and nature of ground cover, and meteorological features such as wind speed and direction and atmospheric stability. Atmospheric stability relates to the extent of vertical mixing in the atmosphere, including, for example, temperature inversions. Release factors include the height of the release point (e.g., at ground level or from the top of a stack), the chemical concentration, velocity and direction of the release, the density and temperature of material released, and other factors. Dispersion modelling is discussed by Crowl and Louvar (2002). Attempts have been made to specify maximum ground level concentrations for chemicals in both emergency situations (typically for exposures of 1-hour duration) and non-emergency situations.

Another frequent concern is that of noise. Excessive noise is important in the occupational health context where prolonged exposure can cause deterioration in hearing capability. Furnaces, high-speed machinery such as centrifuges, gas compressors, and turbines, and intermittent releases of high pressure steam to atmosphere are common sources of high-intensity noise in plants. Adequate ear protection is a key safeguard for operational personnel. Noise nuisance at plant boundaries can depend on intensity at the plant and also on the nature and pitch of the noise, time of day (background noise of traffic and community activities are normally quieter at night), and meteorological conditions including wind direction.

Aqueous effluents to sewers must be treated (if necessary) prior to disposal, and continuously monitored for key properties. Treatment and quality assurance is necessary to protect the sewer pipes from chemical attack and to ensure that feed to downstream purification plant is within specification. Waste water quantities discharged should also be monitored and checked in relation to freshwater consumption and other aqueous inputs and outputs in the process. Storm water effluent may have a separate discharge system to process effluents but should be monitored to check for contamination in order to safeguard the quality of downstream water systems.

Soil and groundwater condition monitoring is important to avoid contamination by toxic chemicals. Groundwater is sometimes used for crop irrigation and as a source of drinking water. Soil contamination is possible through accidental spillages. Solid waste should be monitored for quantity and quality. Where possible, solid waste generation should be minimised using strategies outlined in Chapters 4 to 9. There are regulatory restrictions on the movement of hazardous waste between countries and their regions, encouraging minimisation and treatment of waste at the generating plant.

## 17.7 Emergency Response Planning

Emergency response plans and procedures are important in the context of major safety or environmental accidents. Requirements for emergency response are normally covered in legislation by government safety or environmental authorities. Emergency response plans must involve the whole spectrum of relevant community bodies including police, firefighting, medical support, and media.

Examples of inadequate preparation for emergency response are widespread in the case histories of major safety and environmental disasters. Two relevant cases are the Seveso chemicals release and the Basel warehouse fire, referred to in Chapter 12.

## 17.8 Sustainability Reporting

Process industry companies issue annual reports to shareholders which are in the public domain. Traditionally, these have dealt with management, operational, and particularly financial details relating to a company's performance over a year. Increasingly, process industry companies are also issuing annual sustainability reports, either within the annual report or as a separate dedicated report. Reports vary in content and detail, but indicate a recognition by companies of the importance of reporting on sustainability issues to shareholders and the wider community. Examples of companies issuing separate sustainability reports include

- Alcoa, reporting on its Australian operations which include bauxite mining, alumina refining and aluminium smelting, have published sustainability reports accessible from the company website (see http://www.alcoa.com/australia/en/info_ page/sustain_home.as)

- Santos, a major oil and gas exploration and production venture operating in Australia and internationally, have published sustainability reports on its operations, accessible from its company website (see http://www.santos.com/home.aspx).

Guidelines for sustainability reporting have been drawn up by the Global Reporting Initiative, a non-government body based in Europe (http://www.globalreporting.org).

Company sustainability reports are often a useful source of data on resource consumption and emissions. For example, Alcoa has reported its greenhouse gas emissions for its alumina refining and aluminium smelting operations in Australia.

Sustainability indicators referenced in the Santos Sustainability report for 2009 are detailed in Appendix 1. This report provides examples of projects tackled to improved environmental performance. These examples include reduced flaring, improved energy efficiency in refrigeration compression, improved boiler blow-down control, and water saving initiatives enabling reductions in groundwater extraction.

## 17.9 Emissions Reporting

The National Pollutant Inventory (NPI), established by the Australian Commonwealth Government, was designed to advise the community, government, and industry on types and quantities of specified chemicals emitted to the environment.

Companies using more than specified amounts (commonly 10 tonnes per year) of listed pollutants (other than greenhouse gases) have been required to submit reports. The list of nominated pollutants has been expanded from 36 substances initially to 93 substances in 2010. NPI data come from both industry and non-industry sources. The inventory reflects a recognition by government, industry, and the community of the 'right to know' about emissions. It also enables trends in emissions to be investigated.

The database for NPI reports by substance, source, facility, and location. The estimation technique employed, for example, by direct measurement or mass balance, is also reported. The NPI website address is www.environment.gov.au/net/npi.html.

Many countries are now requiring their larger industrial corporations to report their greenhouse gas emissions. In Australia, under the 2007 National Greenhouse and Energy Reporting (NGER) Act, corporations emitting more than 125 kilotonnes $CO_2$ equivalent per year are required to report their greenhouse gas emissions annually, as direct and indirect emissions. Under NGER 2007, data were reported for the first time in 2010. Further information is available from http://www.climatechange.gov.au/government/initiatives/national-greenhouse-energy-reporting.aspx.

Similar regulatory frameworks are in place for emissions reporting in other countries.

## 17.10  Concluding Remarks

The potential for good project outcomes, as well as safety or environmental damage, is greatest for any project in its operational phase. Ultimately, it is in operations where the vast majority of consumptions and emissions occur and the sustainability of processes and process plants is realised.

Industrial companies are required to monitor and report their emissions under environmental licensing agreements with regulatory bodies, and under emission reporting requirements of government.

EMS and EIPs are important means of identifying environmental goals and improving performance. EIPs also provide special opportunities to consult with stakeholders.

Increasingly, performance in relation to emissions, consumptions, and sustainability indicators are being monitored and reported, and represent an increased commitment to accountability and continuous improvement.

## References

Crowl D. A. and Louvar, J. F. (2002) *Chemical Process Safety*, 2nd edn, Prentice Hall, Upper Saddle River, New Jersey.

Coffey, P. and Donoghue, M. (2006) The Wagerup refinery — beyond the controversy, *Chem. Engg.*, April, 32–36.

Heinsohn, R. J. and Kabel, R. L. (1999) *Sources and Control of Air Pollution*, Prentice Hall, Upper Saddle River, New Jersey.

Lees, F. P. (1996) *Loss Prevention in the Process Industries: Hazard Identification, Assessment and Control*, 3 volumes, 2nd edn, Butterworth-Heinemann, Boston.

Scott, D. and Crawley, F. (1992) *Process Plant design and Operation*, IChemE, Rugby, England.

Townsend, A. (1992) *Maintenance of Process Plant: a Guide to Safe Practice*, 2nd edn, IChemE, Rugby, England.

Turner, B. (1977) Learn from equipment failure, *Hydrocarbon Process.*, November, **56**, 317.

Appendix 1.  Indicators, Santos Sustainability Report, 2009

| Sustainability Indicator | Guidelines | |
| --- | --- | --- |
| | GRI G3 | IPIECA/API |
| **Environment** | | |
| Air quality | ***** | ***** |
| Biodiversity and land disturbance | ***** | ***** |
| Climate change | ***** | ***** |
| Incidents and spills | ***** | ***** |
| Waste management | ***** | ***** |
| Water management | ***** | ***** |
| **Community** | | |
| Community wellbeing | ***** | ***** |
| External stakeholder engagement | | ***** |
| Indigenous rights and cultural heritage | ***** | ***** |
| Product responsibility and reputation | ***** | |
| Social infrastructure | ***** | ***** |
| Transparency and disclosure | ***** | ***** |
| **Our people** | | |
| Governance and Policy | ***** | ***** |
| Health and wellbeing | ***** | ***** |
| Safety | ***** | ***** |
| Workforce capability | ***** | ***** |
| Workforce composition, culture, and commitment | ***** | ***** |
| Workforce remuneration and benefits | ***** | ***** |
| **Economic[1]** | | |
| Financial performance | ***** | ***** |
| Supply chain performance | ***** | ***** |

**Notes**

1. Other economic performance indicators have been detailed in the Santos Annual report for 2009.
2. The symbol **** denotes a related requirement under the guideline source.
3. GRI G3 denotes Global Reporting Initiative (2006), while IPIECA/API denotes the International Petroleum Industry Environmental Conservation Association in conjunction with the American Petroleum Industry.

# Problems: Part D

## 1. Process Flow Sheet Development. Utilisation of By-Product Hydrogen from a Chlor-Alkali Plant

For every tonne of chlorine produced in a chlor-alkali plant some 28 kg of hydrogen are produced. Because of the relatively small quantity of hydrogen produced this potential by-product has often been vented to atmosphere. Better utilisation of by-product hydrogen has been identified by European chlorine manufacturers as one of several key sustainability goals; another key goal is the progressive replacement of mercury cell plants by membrane cell plants.

This assignment explores the use of by-product hydrogen and provides an opportunity to explore how process flow sheets are developed and documented.

(a) Identify two potential uses of hydrogen for each of the following cases:

- internally on the chlor-alkali plant
- externally by other industrial plants.

(b) Consider the case of a hypothetical mercury cell plant of capacity 1000 tonnes per day chlorine.

Assume hydrogen gas at 15 kPa gauge pressure and 90°C leaves the electrochemical reactor saturated with water vapour and mercury vapour. Hydrogen is to be supplied by pipeline to an industrial user, located 3 km away; supply conditions at customer's gate are to be 6 bar g pressure and ambient temperature. Water content is to be lowered to a dew point of 5°C maximum at delivery pressure, and mercury content to 5 μg/Nm$^3$.

(i) Develop a simple block diagram to meet the process needs. Identify the main process steps involved, and the key conditions of process streams leaving each process step. Provide a brief explanation of the reasons for your diagram.

(ii) From your simple block diagram, develop and draw a process flow sheet showing

- flow diagram with all process streams and equipment items
- table detailing the mass flow rate, composition, temperature, and pressure of each process stream
- equipment list identifying all equipment items on the plant.

(c) Estimate by calculation the consumptions of cooling water and electricity on the plant. Specify any other utilities which would be required for use on the operating plant.

(d) How would the process scheme developed differ for the case of hydrogen from a membrane cell plant? Which of the two process schemes would be superior from a sustainability viewpoint and why?

(e) Use an approximation to estimate the diameter of the pipe for hydrogen transport to the user factory. What considerations would govern the choice of pipe diameter in practice?

**Notes**

1. Mercury removal from hydrogen and other gases can be achieved to $5\ \mu g/m^3$ by use of a solid adsorbent of carbon impregnated with sulphur. The removal takes place in a fixed bed of adsorbent, at around ambient temperatures.

2. The following utilities are available:
   Recirculated cooling water: 28°C max, 15°C min, 3 bar g (ground level)
   Electricity: 11 kV/3.3 kV/415 V 3 Ph 50 Hz
   Other utilities can be made available if required and justified.

3. Guidelines for sizing pipes are provided in Chapter 7.

4. The following simplifying assumptions can be made in calculations of

   (a) cooling water consumption — assume gas contains $H_2$ and water vapour

   (b) electricity consumption — consider compression of hydrogen gas alone ignoring water and mercury contents.

5. The following physical property data are required and can be found in *Perrys Chemical Engineers Handbook* (Chapter 3 in sixth edition):

   - vapour pressure of water and mercury
   - density of hydrogen
   - Cp/Cv for hydrogen
   - specific heat of hydrogen and water vapour
   - latent heat of water vapour.

**Considerations in process development in solving problem 1**
As with most design problems, there are a number of possible solutions. The following points should be considered in reviewing various design options.

1. Is mercury adsorbent necessary? Could pressure and temperature conditions for hydrogen gas be selected to achieve mercury removal solely by condensation thus avoiding use of adsorbent? If so, what happens if
   - mist is entrained in gas leaving coolers
   - abnormally high gas temperatures occur because of deficiencies in cooling water, refrigeration, or heat exchanger performance during operation.

2. If mercury adsorbent is necessary, where should adsorbent beds be placed?
   a. At beginning of process?
      Advantage: Eliminates Hg contamination of condensate and down-stream gas
      Disadvantage: Higher mercury load on adsorbent implies greater disposal and/or regeneration frequency
   b. At end of process?
      Advantage: Load on adsorbent would be minimised
      Disadvantage: Upstream condensate would be contaminated with mercury

3. Mercury handling
   a. Could condensed mercury and water be recycled to soda cell? How?
   b. If this is not possible, how could contaminated condensate be handled?
   c. How could spent adsorbent be either regenerated or disposed of securely?

4. Utility consumptions
   Consumptions of cooling water and electricity depend on design decisions, for example,
   a. cooling water temperature rise in heat exchangers
   b. gas temperatures leaving heat exchangers

5. Heat exchanger selection
   What would be the relative advantages of
   a. shell and tube exchanger for gas cooling?
   b. packed column for direct contact gas cooling with water using external cooling of condensate?

6. Pipe sizing
   To what extent is there a trade-off between capital and operating costs?

## 2. Process Flow Sheet Development: Waste Sulphuric Acid

Chlorine-caustic soda manufacture is a major commercial activity in the chemical industry. Chlorine can be handled successfully in mild steel equipment, but wet chlorine is very corrosive to most materials. Wet chlorine can be dried successfully using sulphuric acid, but often it is difficult to find a use or market for dilute sulphuric acid produced. This problem focuses on

- development of a process flow sheet for chlorine purification
- accounting for process resources used and wastes, and utilities used
- exploring options for treatment of some process effluents.

### Problem

A chlorine caustic soda plant of 200 tonnes/day chlorine capacity uses membrane cells to convert sodium chloride brine to chlorine, caustic soda and hydrogen. Chlorine gas leaves the cells saturated with water vapour at 20 kPa gauge pressure and 90°C. The gas composition by volume on a dry basis is 98% $Cl_2$, 1.8% $O_2$, and 0.2% $H_2$. The gas is to be cooled, dried using sulphuric acid, and compressed by pipeline supply to a user plant on the same site. The user plant requires chlorine supplied at 100 kPag and a temperature not exceeding 40°C. The maximum moisture content of the chlorine gas after drying is 0.01 g/$Nm^3$ (N indicates 0°C and 1 bar abs). The site can be assumed flat. The following utilities and raw materials are available to the plant.

| | Cost |
|---|---|
| Towns water: Mains pressure | $ 1.00/$m^3$ |
| Cooling water: 28°C max., 15°C min., 3 bar g (ground level) | $0.07/$m^3$ |
| Chilled water: 5°C, 3 bar g (ground level) | $0.35/$m^3$ |
| Steam: Medium pressure 18 bar g saturated. | $15/tonne |
| Electricity: 11 kV / 3.3 kV / 415V 3Ph 50 Hz | $80/MWh |
| Sulphuric acid: 98% by mass, supplied to battery limits | $80/tonne |
| Caustic soda: 32% by mass from cells | At cost |

### Tasks

(a) Develop and document a process flow sheet for a plant to cool, dry, and compress the chlorine gas leaving the cells. Document the process flow sheet to provide a detailed table of the process raw materials consumed and the process waste streams emitted. Include mass flow rates, compositions, temperatures, and pressures.

(b) Calculate and list consumptions of electricity, cooling water, and chilled water (if used) in the process, and identify any other utilities required.

(c) Recommend a means of treating and reusing the water condensed from the gas stream as a result of cooling the chlorine gas stream.

(d) (i) Recommend a method for reprocessing the diluted waste sulphuric acid for reuse in the drying process. Comment on the viability of the method.

   (ii) Recommend a possible market for the waste sulphuric external to the chlorine plant.

   (iii) Recommend a means of recycling the waste sulphuric acid external to the chlorine plant should options (i) and (ii) above not be viable.

## Notes

1. A solid chlorine hydrate forms at $14°C$.
2. Sulphuric acid drying systems for chlorine plants normally operate in either two or three stages with corresponding acid effluent streams of $\sim78\%$ w/w $H_2SO_4$ and $\sim54\%$ w/w $H_2SO_4$.

## References

Ullmann (2002) *Encyclopaedia of Industrial Chemistry*, Wiley-VCH, Weinheim, Germany.

Perry, R.H. and Green, D. ed. (1984) Perry's Chemical Engineers Handbook (Refer Table 3.5 Vapour pressure of water, and Table 3-13 Water partial pressures over aqueous $H_2SO_4$ solutions) (6th ed.) McGraw-Hill, New York.

## 3. Process Flow Sheeting: Carbon Dioxide Separation from Flue Gas

Carbon dioxide sequestration has been suggested as a means of reducing emissions of carbon dioxide to atmosphere. For an existing power station using fossil fuels, this implies separating the carbon dioxide from flue gas and transporting it by pipeline to the point of sequestration. Victorian power stations based on brown coal are located in the Latrobe Valley and provide the large majority of Victoria's electricity generating capacity. Some existing and proposed power stations use natural gas sourced from offshore Victoria.

(a) Specify the mass flow rate, composition, temperature, and pressure of the flue gas leaving a power station of 500 MW capacity and using

(i) natural gas
(i) brown coal.

The mass flow rate and composition of the flue gas should be calculated making simplifying assumptions (see later). Flue gas temperature and pressure should be specified based on reasoned estimates, considering temperature driving forces, flue gas dew point, and pressure and temperature losses.

(b) Specify the pressure, temperature, and composition of the carbon dioxide at the inlet to the transportation pipeline, based on reasoned estimates.

(c) Develop and draw a process flow sheet for a plant to treat flue gas from the natural gas power station to separate, recover, compress, and cool $CO_2$ for delivery to the sequestration site. Terminal points of the flow sheet are

- flue gas stream leaving the power station after steam has been generated and any particulates have been removed
- treated gas entering the pipeline for delivery to the sequestration site.

The first stage of the flow sheet development should be a simplified block diagram showing the main process steps. For the $CO_2$ separation step, absorption stripping has been most widely used, but other alternatives should be identified and their advantages and disadvantages considered. The detailed process flow sheet document should comprise two parts:

- a process flow diagram showing major equipment items
- a table of process streams detailing mass flow rates, temperatures, pressures, and component compositions.

The detailed process flow sheet should be accompanied by

- a concise statement of the process design basis for the flow sheet
- a clear and concise presentation of mass and energy balance calculations for the feed gas and the $CO_2$ separation, compression, and cooling steps.

(d) What modifications, if any, to the process flow sheet would be required in order to capture and sequester $CO_2$ from flue gas derived from the power station fired by brown coal? Answer qualitatively, aided by a block diagram.

(e) Make some conclusions about the sustainability of sequestering $CO_2$ from power stations based on the above work.

## Data

Details of fuel compositions are provided below.

### Natural gas composition

| | | |
|---|---|---|
| $CH_4$ | %vol | 90.6 |
| $C_2H_6$ | %vol | 5.6 |
| $C_3H_8$ | %vol | 0.8 |
| $C_4H_{10}$ | %vol | 0.2 |
| $N_2$ | %vol | 1.1 |
| $CO_2$ | %vol | 1.7 |
| Sulphur | $mg/Nm^3$ | 45 |
| Water | $mg/Nm^3$ | 60 |
| Lower CV (calorific value) | MJ/kg | 46.7 |

### Brown coal composition

| | | |
|---|---|---|
| Moisture | mass % | 62 |
| Carbon | mass % | 25.3 |
| Hydrogen | mass % | 1.9 |
| Sulphur | mass % | 0.11 |
| Nitrogen | mass % | 0.23 |
| Ash | mass % | 0.8 |
| Oxygen | mass % | 9.7 |
| Lower CV (wet basis) | MJ/kg | 8.4 |

### Simplifying assumptions
The following simplifying assumptions are suggested for calculations.

1. All carbon and hydrogen in the fuel are fully combusted.
2. Combustion air is free of moisture and $CO_2$ and comprises 79% $N_2$, 21% $O_2$ by volume, and 76.7% $N_2$ and 23.3 % $O_2$ by mass.
3. $NO_x$ (nitric oxide and nitrogen dioxide), $N_2O$, and $SO_2$ will be produced in fuel combustion but quantities need not be calculated.
4. Power cycles, power generation efficiency, excess air for fuel combustion are as follows:

| Fuel | Power Cycle | Efficiency* | Excess Air |
|------|-------------|-------------|------------|
| Brown coal | Steam turbine | 30% | 25% |
| Natural gas | Combined cycle gas turbine | 50% | 10% |

$$* \text{ Efficiency} = \frac{\text{electric power generated}}{\text{fuel energy consumed}}$$

# Index of Topics

Abiotic resources  207, 208
Abnormal operation  173, 185, 201,
    255–258
Accident
    investigation  256–257
    prevention  255–256
Added value  309
Adsorption  90–91, 101, 166, 178
Agent materials  24, 55, 67, 101
Agitation in vessels  147–149
Air cooling  124
Allocation  202, 203
Ammonium phosphate  39, 40, 199, 331
Aqueous effluent  63, 85, 88, 103, 123,
    165, 166, 377
Atom utilisation  58

Basel accident  254, 378
Bath-tub curve  371
Battery limits  268, 284, 358
Beer packaging  49, 161
Bhopal accident  248
Biomass  16, 330
Biotic resources  207, 208
Boiler  106, 113, 119, 127
    blow-down  115
    water quality  115, 119, 120
Brundtland report  6, 9–10
Burden-benefit analysis  299–307
By-products  60, 70, 219, 331
By-product hydrogen  38, 66, 68–71,
    338

Capital investment  29, 265–273, 290,
    358
Capital recovery  276–277
Carbon dioxide  10, 22, 109–110,
    167–169, 206–207, 210–211,
    218–219, 315–317, 319–322
    capture from flue gas  307, 387–390
Cash flow  282–289, 359
Catalysis  65–67, 73, 75, 172, 235, 327,
    336
Centrifuge  79, 92, 130
Chernobyl  253–254
Chlorine
    manufacture  68–71
    purification  386
Classification
    costs  275, 278
    within LCA  205–214
Cleaner production  19–30
Cleaning  174–175, 354
Climate change  6, 16, 17, 196,
    206–207, 292, 381
Club of Rome  4
Coal  105, 107, 109, 110, 126–128, 180
    classification  109
    composition  109, 315, 389
Cogeneration  129
Combustion  103–107, 110–112, 169,
    186–187, 235, 242
Commodity chemicals  280
Compressed air  130
Compressor  134–139, 183
Conversion  55–59, 64, 67, 71–76, 79
Coode Island incident  254
Cooling towers  119–123

Cooling water  119–124, 187, 230–231, 257
  circuit  121
  cost  277–278
  quality  120
  temperature rise  122–123, 385
Copenhagen conference (2009)  6, 17

Deaerator  113–114
Deepwater Horizon Drilling Rig  254
Depreciation  276, 282
Design process  345–351
Diesel hydrotreating  233–244, 302–305
Distillation  85–87, 88–89, 100–101
  azeotropic  85, 101
Drying  93, 96–97, 183, 336, 386–387

Earth Summit, Rio de Janeiro  11
Eco-indicator  225
Economies of scale  289–291, 307–308
Economiser  113–114
Electricity generation  126–129, 230–233, 304–307, 320–322
  use  129
Emergency response planning  261, 378
Emission taxes and trading schemes  293, 340–341
End use analysis  280
Energy recovery  63, 72, 74, 76–77, 116, 117–118, 149–150
Enviro-economic assessment  297–308
Environmental externalities  291–293
Environmental improvement
  plan  374–375
Environmental legislation  3–6, 340–341
Environmental management costs  278
Environmental management
  sytems  372–374
Environmental performance
  Monitoring  376–378
  Reporting  378–381

Environmental Priority
  Strategies  224–225
Environmental projects  28–29, 265–266, 281–282
Equipment costs  270–271
  design  97–99, 134–151, 346, 348
Equity funds  267, 277
  intergenerational  10, 14, 16
  intragenerational  10–12
Ethylene manufacture  71–73, 125, 155–156
Ethylene markets  280, 281
Evaporation  93–94
Event tree  256, 257
Externalities  291–293
Exxon Valdez  35, 177, 255
Explosion  248, 250, 251, 254
Extraction
  liquid–liquid  95
  solid–liquid  94–95
Extraneous materials *see* Agent
  materials

Failure tolerance  256, 262–263
Fault tree  256, 257, 258
Feedstock
  impurities  60–61
  purification  69, 75
Filtration  60, 69, 70, 91–92
Fine chemicals  280
Fires  250, 254
Fixed capital  267–273
  inside battery limits  268
  off-sites  268
Fixed operating costs  274–276
Flammability Limits  250, 251, 264
Flare stacks  112–113
Flixboro accident  248
Flue gas  106–107, 109–112
Forecasting  280–281, 325–327, 357–359
Fouling  175, 176
Fuel  105–109
  combustion *see under* combustion
  combustion burners  111, 112
Fuel ethanol  46, 47, 100–101

Fugitive emissions  179–180, 218, 242, 354
Fusel oils  101

Gas absorption  87–89
  cases  97–100
  column internals  97–99
Gate to gate  21, 198
Glass bottle weight reduction  161
Glass recycling  163–164
Global warming  3, 206–208
Green engineering  6–7

Hazard identification  255
HAZOP studies  248, 255, 256, 262–263
Heat exchangers  145, 146
Hydrogen
  car fuel  47
  refinery  38, 237
  *see also* By-product hydrogen
Hydrogen cyanide  73, 336

Impact categories  206–213, 239–244, 315–318
Integration
  of criteria assessments  362–364
  of plants and industries  333–334
Isenthalpic expansion  117
Isentropic expansion  116–117
Incineration  168–169
Industrial ecology  31–37
  examples  38–42
  in industry planning  333–335
Inert gas  130
Inflation of capital costs  269, 295
Inherent safety
  application  363
  principles  259, 263
  quantification  259
Insulation  150
Intensification  327, 328
Integrated site manufacture  36, 37

Johannesburg summit  6, 11

Kalundborg symbiosis  41
Kwinana symbiosis  41
Kyoto protocol  6

Land degradation  167–168, 207, 228
Landfill  167–168, 292–293
Life cycle assessment (LCA)  193, 229
  goal definition  196, 197, 217
  impact assessment  7, 196, 210, 221, 225, 226, 239, 298, 363
  improvement analysis  196, 215
  interpretation  196, 349
  inventory analysis  196, 201–205
  inventory data for utilities  229–233
  scope of study  196–197
  system boundary definition  25, 26, 197–200, 234
Life cycle assessment cases
  Diesel desulfurisation  233–244
  Electricity generation  231–233
  $SO_2$ emissions from smelters  198–217
Life cycle assessment software  225
Life cycle costs  293–294, 358
Lifestyle implications  341–342

Maintenance  174–175, 370–372
Maintenance costs  276
Manufacture
  at point of use  334, 340
  centralized vs distributed  334, 335
Material flow through economy  31
Materials recycling  22, 64, 67, 155, 160, 188
Materials transport  40, 176–177, 218, 331, 332
Mercury
  in chlorine manufacture  68–71, 383–385
  in fossil fuels, minerals  71
Mining, extraction of resources  213, 220

Mist formation and removal 92–93
Mixing
  in agitated vessels 62, 147–149
  of reactants 62
Monitoring
  process streams 171–172, 354, 377
  emissions 6, 376–380
Montana oil field leak 254
Montreal protocol 6

Natural gas
  in electricity production 126–128
  purification 316
  sales gas composition 107
  wellhead gas 107–108
Nitrogen dioxide 109–111, 212
Nitrous oxide 206, 212
Noise 207, 377
Normalisation in LCA 215, 221, 223
Nutrification (*also* eutrophication)
  208, 209

Oil rig accidents 254
Oil spillage from ships 177, 178, 255
Operating costs 274–280, 287
Operations 171, 369
Ozone depletion 208, 209, 211, 212

Packed columns 87, 97, 98–99, 147
Pay back time 283, 289
Performance curves
  compressors 135–136
  pumps 142–143
Personnel costs 268–272, 274–276
Phosphate rock 39, 60
Phosphogypsum 41, 60
Photochemical smog 207–208
Pipe sizing 144, 384, 385
Piper Alpha incident 251
Piping and instrumentation
  diagram 354
Plant layout 151, 251
Plant shut down 175, 370
Plant start up 171, 172

Plastics recycling 161–163
Pollution prevention 20
Pollutant
  transport paths 69
Polyvinyl chloride (PVC) 78–80
Power consumption 137, 139, 143, 148
Pressure loss
  in piping 123, 144
  in equipment 145–147
Process control 133, 171, 173, 354
Process flowsheet
  development 349, 351
  documentation 353, 365
Product life cycle 20, 45, 47, 159, 193,
  319
Profitability estimation 282, 289, 310
Project development 354–356, 358,
  359, 363, 364
Project management 362
Project sanction 355
Pumps 139–144, 257

Raw material
  *see under* Feedstock
Reaction chemistry 55, 56, 58, 63
Reactors 55, 56, 57, 62, 70, 338
Recycling 64, 155–165
  closed and open loop 157
  on-site and off-site 159
Refrigeration 124–126
  refrigerant selection 125
Renewable energy 105, 129
Renewable feedstocks 329, 330
Resource depletion 208, 210, 213, 215,
  220, 221
Responsible care 375–376
Revenue 280, 286
Reversible reactions 64–65
Risk 173, 256–261
  assessment 257–258
  perception 310–312
Root cause analysis 249

Sandestin principles 7
Scales in industrial ecology
  macro, meso, micro 31–37

Scenario postulation and assessment  13, 15, 313, 326, 358
Screening of alternatives  350
Sea water cooling  123–124
Secondary coolants  123
Secondary reactions  63, 67, 71, 74
Selectivity  59, 62
Separation processes  83, 156
Seveso accident  253
'Silent Spring'  3
Site integrated manufacture  36–37
Site selection  177, 331–333
Spanish campsite disaster  252
Specialty chemicals  280
Staged assessment and review  363
Stakeholder engagement  13, 341
Static mixers  62, 149
Steam generation  113–116
Steam use  116–119
Storm water  180, 377
Sulphur flow through economy  34
Sulphur dioxide  75, 197, 198, 212
Sulphuric acid
  in chlorine drying  96–97
  manufacture  75–78, 150, 198, 300–302
Superheater  113, 117
Sustainability  9–17
  assessment  297
  concepts  9
  indicators  13, 308, 310–311
  reporting  378–381
  time horizons  2, 14
System boundary definition  25, 200
  cradle to grave  21, 198
  cradle to gate  21, 198
  gate to gate  21, 198

Tax, carbon  293
  corporate  282, 285, 286
  property  276, 285
Technology
  diffusion  328
  evolution  328
  innovation  327
Theoretical flame temperature  111
Toxic release  4, 207–208, 248, 250, 253
Toxicity
  human  207, 208, 209, 212, 223
Trade-offs  87, 299–307, 362
Transport of materials  35–36, 176–179, 218
Tray columns  86, 98–99
Triple bottom line  12

Utilities  103–132, 229–233

Vacuum  131
Variable costs  275, 290, 291
Variable speed drives  143

Waste classification  27–28
Waste minimisation  23, 53–80, 83–102, 171–181
Waste treatment  164–166, 169, 198
Water  15–16, 119
Water recycling  165–166
Weighting of components in impact assessment  210, 224
Working capital  267, 273, 286

Yield  59, 72, 73

# Index of Cases

Basel incident  Ch 12
Bhopal incident  Ch 12

Chernobyl incident  Ch 12
Chlorine-caustic soda manufacture
    Ch 4
Chlorine-caustic soda manufacture,
    costs and profit evaluation  Ch 13
Chlorine drying  Ch 5
Coode Island incident  Ch 12

Diesel hydrotreating  Ch 11, Ch 14

Electric power generation from fossil
    fuels  Ch 11, Ch 14
Ethanol distillation  Ch 5
Ethylene manufacture  Ch 4

Fertiliser complex, Queensland  Ch 3
Flixboro incident  Ch 12

Glass recycling  Ch 8

Hydrogen cyanide manufacture  Ch 4,
    Ch 15
Hydrogen utilisation in refineries
    Ch 3

Kalundborg complex  Ch 3

Landfill  Ch 8, Ch 13

Plastics recycling  Ch 8
Polyvinyl chloride manufacture
    Ch 4

Seveso incident  Ch 12
Sodium cyanide manufacture  Ch 4,
    Ch 5, Ch 15
Sulphur dioxide treatment  Ch 10
Sulphuric acid manufacture  Ch 4,
    Ch 14

Utilities, life cycle inventories  Ch 11

# Index of Set Problems

Beer packaging  Part A, Problem 5

Carbon dioxide separation from flue
    gas, flowsheet development
    Part D, Problem 3
Chlorine production and
    purification  Part B, Problem 1

Electricity generation
    LCA  Part C, Problem 1
    Sustainability  Part C, Problem 5

Fuel ethanol  Part A, Problem 2

Hydrogen as fuel, sustainability  Part A,
    Problem 3
Hydrogen utilisation, flowsheet
    development  Part D, Problem 1

Materials recycling  Part B, Problem 6

Natural gas purification LCA  Part C,
    Problem 2
Nitric acid manufacture, industrial
    ecology  Part A, Problem 4
Nitric acid manufacture, waste
    minimisation  Part B, Problem 3

PVC product sustainability  Part A,
    Problem 1
PVC production process  Part B,
    Problem 2
PVC production, water consumption
    and $CO_2$ emissions  Part C,
    Problem 4

Recirculated cooling water  Part B,
    Problem 5

Sea water desalination, project
    evaluation  Part C, Problem 5
Steam generation  Part B, Problem 4
Sulphuric acid, waste minimisation
    Part D, Problem 4

Milton Keynes UK
Ingram Content Group UK Ltd.
UKHW031136141024
449569UK00006B/148